William A. Campbell

Outlines of Anatomy for Students

A Guide to Dissection Based on Morris's Text Book of Anatomy

William A. Campbell

Outlines of Anatomy for Students
A Guide to Dissection Based on Morris's Text Book of Anatomy

ISBN/EAN: 9783337249854

Printed in Europe, USA, Canada, Australia, Japan

Cover: Foto ©berggeist007 / pixelio.de

More available books at **www.hansebooks.com**

FOR

STUDENTS

A GUIDE TO DISSECTION

BASED ON MORRIS'S TEXT-BOOK OF ANATOMY.

BY

WILLIAM A. CAMPBELL, M. D.,

DEMONSTRATOR OF ANATOMY IN THE MEDICAL DEPARTMENT OF THE UNIVERSITY OF MICHIGAN.

PHILADELPHIA:

P. BLAKISTON, SON & CO.,

1012 WALNUT STREET.

1895.

PRESS OF WM. F. FELL & CO.,
1220-24 SANSOM ST.,
PHILADELPHIA.

CONTENTS.

TO THE STUDENT.

In this outline the numbers refer to the pages and plates in Morris's Anatomy, dated 1895. (The numbers in parentheses refer to the edition of Morris dated 1893. This edition is identical with that dated 1895, with the exception of the paging.) Looking on the page referred to, it will be seen that the corresponding heading is printed in heavy-faced type, and the paragraph, or portion of paragraph following it, should be read. The structures of each region should be studied successively, *in the order indicated by the outline.* The student should carefully work out each structure, studying each part fully as it is exposed, comparing with the description in Morris. All structures should be carefully *exposed in place* in order that the *relations* may be properly studied and no part should be cut or removed except as indicated. When a heading is inclosed in a parenthesis it is to indicate that the structure mentioned cannot be fully exposed at that point in the work, but will be shown at another period of the dissection.

OUTLINES OF ANATOMY.

THE HEAD, AND ANTERIOR CERVICAL STRUCTURES.

Superficial Anatomy of the Head and Face.
Bony landmarks, 1111–12–13. (1088–9–90)
    Lambda. Bregma. Pterion.
The bony sinuses, 1113–14–15. (1091–2)
    The frontal.
    Mastoid sinuses.
    Sphenoidal sinuses.
Arteries, 1117–18. (1094–5–6)
    Supraorbital. Temporal. Occipital. Posterior auricular.
    External carotid. Facial. Frontal artery.
Parotid duct. Sheath of the parotid, 1119. (1096)
Position of the lachrymal puncta, lachrymal sac, 1120–1 ; Fig. 667. (1097)
    Manipulation of a probe along the lachrymal passages.
The Mouth, 1121–2–4; Fig. 668. (1098–9–1101)
    Wharton's ducts. Frænum. Sublingual glands.
    Submaxillary gland. Genio-hyo-glossi. Lingual nerve.
    Coronoid process. Hamular process of the sphenoid.
    Tonsils :—
        Relations of the tonsils.
    Finger introduced downward at the back of the mouth.
    If the finger be moved upward. Size of the nares.
    The palate :—
        Isthmus faucium.
The Nose, 926. (905–6)
    Nose proper :—
        Root. Bridge or dorsum. Top or lobe. Base.
        Nostrils. Columna. Vestibule.
        Alæ or wings.
        Bony framework.
        Cartilaginous portion.
The Scalp, 1115. (1092)
    Layers: Fig. 663.
        (1) Skin. (2) Subcutaneous fat and fibrous tissue.
        (3) Occipito-frontalis and aponeurosis. (4) Subaponeurotic layer. (5) Pericranium and subpericranial connective tissue.

Dissection of the Scalp.
Skin, hairs, sebaceous glands, 1115 ; Fig. 663. (1092)

Carefully remove the skin and expose (2) the layer of subcutaneous fat and fibrous tissue, in which will be found the superficial nerves and vessels. To remove the skin, make a median incision from the root of the nose to the external occipital protuberance, and a transverse incision from the lower end of the mastoid process on one side, over the vertex to the corresponding point on the opposite side. Connect the ends of these incisions by horizontal incisions on each side, as follows: Beginning in front at the root of the nose, make an incision along the superior border of the orbit to the outer angle of the

2

eye and backward to the tragus of the ear, then over the ear to the transverse incision. Also, an incision from the lower end of the mastoid process backward to the occipital protuberance, connecting the ends of the transverse and median incisions. Begin at the vertex and turn the skin down and off. Much care must be exercised to turn off the skin only, as the three outer layers of the scalp are closely connected together. The nerves and vessels should then be carefully exposed.

The superficial fascia, 452. (447)
Subcutaneous fat, (2) 1115. (1092)
Superficial nerves, Fig. 449 ; P. 809. (791)
    Branches of the frontal, 796. (778)
           Supraorbital nerve.
           Supratrochlear nerve.
    Temporal branch—of the orbital or temporo-malar nerve, 798. (780-1)
    Temporal branch—of the facial nerve, 810. (791)
    Auriculo-temporal nerve, 803. (785)
        Anterior auricular branches.
        Superficial temporal branches.
    Posterior auricular nerve, 808. (790)
    Great auricular nerve, 830. (810-11)
           Mastoid branch.   Auricular branches.
    Lesser occipital nerve, 829-30. (810)
           Auricular branch.   Mastoid branch.   Occipital branches.
    Great occipital nerve, 827. (808)
Vessels of the scalp, 1116. (1093)
Arteries :—
    Frontal branch, 533 ; Fig. 334. (524)
    Supraorbital artery, 531. (522)
    Temporal artery or superficial temporal artery, 521-2. (513)
        Branches of the temporal artery :—
           Middle temporal branch or middle deep temporal artery.
           Orbital or zygomatico-orbital branch.
           Anterior terminal branch, secondary branches.
           Posterior terminal branch.
    Posterior auricular artery, 520-1 ; Fig. 335. (512)
        Anterior terminal or auricular branch.
        Posterior terminal, or mastoid or occipital branch.
    Occipital artery—third part of its course, 519 ; Fig. 335. (511)
           Terminal or superficial branches, 520. (512)
Veins of the head and neck, 648. (635-6)
    Superficial and deep.
        Superficial.
        Deep veins.
    Superficial veins of the scalp and face, 648 ; Fig. 385. (636)
        Anterior.   Posterior.   Lateral.
        Anterior vein, 648-9-50. (636)
           Frontal vein.
             Tributaries.
                Supraorbital vein.
        Lateral veins, 652. (639-40)
           Anterior lateral veins :—
             Superficial temporal.
             Middle temporal vein.
                Orbital branch.
        Posterior lateral veins, 653. (640-1)
             Posterior auricular vein.
                Tributaries.
        Posterior vein, 651-2. (639)
             Occipital vein.
Emissary veins, 1116. (1093)
Lymphatics, 688. (673)

Lymphatics of the head and neck :—
    Superficial. Deep.
 Superficial lymphatic vessels of the scalp, 689 ; Fig. 398. (674)
 Superficial lymphatic glands of the head and neck, 689 ; Fig. 398. (674)
  Transverse :—
   Occipital or suboccipital glands.
   Posterior auricular or sterno-mastoid glands.
Muscles of the head, Fig. 309 ; P. 453. (448)
 Occipito-frontalis, 454–5. (449–50)
  Occipitalis and frontalis and epicranial aponeurosis.
   Occipitalis :—Origin. Insertion.
   Frontalis :—Origin. Insertion.
   Epicranial aponeurosis.
   Structure. Nerve supply, occipitalis, frontalis.
   Action. Relations. Variations.

Make a median incision three or four inches in length through the aponeurosis of the occipito-frontalis, also a short lateral incision. Raise a small portion of the aponeurosis and expose the subaponeurotic layer.

Subaponeurotic layer, 1115. (1092–3)
Pericranium and subpericranial connective tissue. 1115. (1092)
Extrinsic Muscles of the Auricle, 458–9 ; Fig. 309. (452–3)

To demonstrate these muscles, draw the auricle in a direction from the point of origin, when the fibers will be rendered prominent and can be exposed.

 Attollens aurem :—
  Origin. Insertion. Structure. Nerve-supply. Action. Relations.
 Attrahens aurem :—
  Origin. Insertion. Structure. Nerve-supply. Action. Relations.
 Retrahens aurem :—
  Origin. Insertion. Structure. Nerve-supply. Action. Relations.

With a chisel chip off a small portion of the outer table of the skull cap and examine the diploë.

Structure of the Cranium, 1116–17. (1093–4)
 Two layers and intervening cancellous tissue.
  Outer. Inner.
  Diploë.
   Veins of the diploë, 655–6 ; Fig. 386. (642–3)
    Frontal. Fronto-sphenoidal
    Fronto-parietal or anterior temporal.
    External parietal or posterior temporal.
    Occipital or parieto-occipital.
  Results of the above varying elasticity, 1016–17. (1093–4)
Anatomical conditions tending to minimize the effects of violence inflicted upon the skull, 1117. (1094)
  (1) Density and mobility of the scalp.
  (2) Dome-like shape of the skull.
  (3) Number of bones.
  (4) Sutures.
  (5) Internal membrane.
  (6) Elasticity of outer table.
  (7) Overlapping of some bones.
  (8) Presence of ribs or groins.
  (9) Buttresses.
  (10) Mobility of the head upon the spine.

Remove the skull-cap in the manner indicated, Dissection, 710. (694)

The Cranial Cavity.
The meninges, 710. (694)
 Dura mater, arachnoid, and pia mater.
Dura mater, 711. (695)
 Outer or periosteal lamina.

Dura mater :—
   Inner or supporting lamina.
Pacchionian glands, 716. (700)
Deep veins of the head and neck, 655. (642)
A cranial sinus, 711. (695)
   Venous sinuses of the cranium, 656–7. (643)

Open the superior longitudinal sinus by a median incision ; trace it from the foramen cæcum to the internal occipital protuberance.

Superior longitudinal sinus, 657–8. (643–4–5)
   Torcular Herophili.

Turn off the lateral portions of the dura mater in the manner indicated, and note the cerebral veins, Dissection, 711. (694–5)

Subdural space, 715. (699)
Veins of the brain, 661. (648)
   Cerebral veins.
   Cortical, hemispherical, or superficial veins.
      Superior cortical veins.
Falx Cerebri, 714. (698)

Divide the falx cerebri in front and throw it backward and out of the longitudinal fissure, and carefully remove the brain ; to do this, tilt the head back, carefully raise the frontal and olfactory lobes from the anterior fossa, divide the optic nerves and the ophthalmic arteries just back of the optic foramen. Divide the internal carotid arteries also. In the median line behind the optic commissure, the pituitary body lies in the sella turcica, covered by the dura mater, but connected with the brain by the infundibulum. Divide the infundibulum close to the dura mater. The third pair of nerves will next be exposed, and should be carefully divided midway between the brain and the dura mater. Proceed thus, dividing each pair of cranial nerves as they are displayed. Divide the tentorium near its attached border and push it back ; the medulla and cerebellum can then be raised. Cut the spinal cord low down in the canal and divide the vertebral arteries where they emerge from the dura mater. The brain can then be removed, and should be placed in spirit or Müller's fluid and properly hardened before dissection. The dura mater lining the base of the cranium should then be examined. The cranial nerves should be exposed and traced to place of exit from the cranial cavity. Each of the venous sinuses should be laid open with the point of the knife and their communications noted. Finally, the pituitary should be exposed and removed and its structure studied.

The dura mater, diaphragma sellæ, 712. (696)
Tentorium Cerebelli, 714–15. (698–9)
   Upper surface. Under surface. Free border. Attached border.
Falx Cerebelli, 715. (699)
Diaphragma sellæ, 715. (699)
The Cranial Nerves, 786–7–8. (769–70)
   Superficial and deep origins.
   General distribution :—
      Nerves of special sense.
      Motor nerves.
      Mixed.

In the following list the place of exit of each cranial nerve is stated. The superficial and deep origins will be studied later. A statement of the course of each nerve will be found in the general description, to which the reference is given. The distribution of each nerve will be gradually developed as the dissection proceeds.

(Modified from Holden.)

The Cranial Nerves :—
   First, or olfactory nerve, 788. (770)
      From the under aspect of the bulb proceed about twenty branches, which pass through foramina in the cribriform plate of the ethmoid bone.
   Second, or optic nerve, 790. (772)
      Passes through the optic foramen into the orbit, accompanied by ophthalmic artery.
   Third, or oculo-motor nerve, 791. (774)
      Passes through the dura mater close behind the anterior clinoid process, traverses the outer wall of the cavernous sinus, and enters the orbit through the sphenoidal fissure.

The Cranial Nerves:—
  Fourth, or trochlear nerve, 793. (775)
    Passes through the dura mater a little behind the posterior clinoid process.
    It passes through the outer wall of the cavernous sinus, lying below the
    third nerve and above the first division of the fifth, and then runs forward
    through the sphenoidal fissure, passing above the third nerve.
  Fifth, or trigeminal nerve, 794. (776)
    The fifth nerve passes through an aperture in the dura mater under the free
    border of the tentorium at the apex of the petrous portion of the temporal
    bone. It consists of two parts—a larger or sensory root, and a smaller or
    motor. Upon its sensory root is developed a large ganglion, the Gasserian
    ganglion ; the motor root passes below and is not connected with it. From
    this ganglion proceed the three primary divisions of the nerve, the ophthal-
    mic, maxillary, and mandibular.
    Gasserian ganglion, 795. (777)
        Ophthalmic division of the fifth nerve, 795. (777)
        Maxillary division of the fifth nerve, 797. (780)
        Mandibular division, 801. (783)
  Sixth, or abducent nerve, 805. (787)
    The sixth nerve pierces the dura mater behind the body of the sphenoid
    bone, which it grooves. It then passes along the inner wall of the caver-
    nous sinus, external to the internal carotid artery, and enters the orbit
    through the sphenoidal fissure.
  Seventh, or facial nerve, 806. (788)
    Passes outward through the meatus auditorius internus.
  Eighth, or auditory nerve, 811. (792)
    Passes outward through the internal auditory meatus with the facial. It is
    the larger of the two nerves, and lies below the facial.
  Ninth, or glosso-pharyngeal nerve, 813. (794)
  Tenth, or pneumogastric nerve, 815. (796)
  Eleventh, or spinal accessory nerve, 819. (800)
    The three nerves pass through the jugular foramen, the ninth in front of the
    pneumogastric, the eleventh behind it.
  Twelfth, or hypoglossal nerve, 820. (801)
    Passes through the anterior condyloid foramen.

The sinuses should now be demonstrated. Trace each, laying it open with the point of the knife.

Venous sinuses of the cranium, 656–7–8–9–60–1. (643–5–6–7–8)
  Inferior longitudinal or inferior sagittal sinus.
  Straight sinus, or sinus rectus. Torcular Herophili.
  Occipital sinus, marginal sinuses.
  Lateral sinus. Transverse sinus. Sinus jugularis. Sigmoid sinus.
    Right lateral sinus. Left.
  Cavernous sinus :—
    In front. Internally. Posteriorly.
  Circular sinus, or intercavernous sinuses.
                  Anterior, posterior, inferior, etc.
  Superior petrosal sinus.
  Inferior petrosal sinus.
  Transverse or basilar sinus.
  Spheno-parietal sinus, or sinus alæ parvæ.
  Petro-squamous sinus.

  The cranial sinuses, 711. (695)
      Larger system :—
          Superior longitudinal, inferior longitudinal, straight, occipital, petrosal.
          Torcular Herophili. Lateral sinuses.
      Smaller system :—
          Spheno-parietal, cavernous, circular, transverse, inferior petrosal sinuses.
      Emissary veins, 711, 1116. (695, 1093)

Intercranial portion of the internal carotid, 529. (520)
    Branches of the intercranial portion, 530. (521)
        Arteria receptaculi.
        Pituitary branches.
        Gasserian or ganglionic branches.
        Meningeal or anterior meningeal branches.
Middle or great meningeal, 524. (515-16)
        Anterior branch.
        Posterior branch.
        Gasserian branches.
        Petrosal branch.
        Tympanic branch.
        Orbital or lachrymal branch.
            Anastomotic or perforating branches.
Meningeal arteries, 713. (696-7)

  In the anterior cranial fossa :—

      Anterior meningeal branches.
      Twigs—from middle meningeal.

  In the middle cranial fossa :—

      Branch—of ascending pharyngeal.
      Meningea parva.
      Meningeal branch of internal carotid.

  In the posterior cranial fossa :—

      Meningeal branches—from occipital.
      Twigs—from occipital and ascending pharyngeal.
      Meningeal branch—of the vertebral.

Lymphatics from the interior of the cranium, 691. (676)
    Meningeal.
    Cerebral lymphatics.

Carefully raise and throw forward the Gasserian ganglion and expose the petrosal nerves.

Great superficial petrosal nerve, 800, 808. (782, 789)
Lesser superficial petrosal nerve, 808. (789)
External superficial petrosal, 808. (789-90)
Carotid plexus, 865. (845)
      Tympanic branch.
      Great deep petrosal nerve.
      Branches to the Gasserian ganglion.
      Branches to the sixth nerve.
Cavernous plexus, 865. (845)
      Communicating branches to the third, fourth, and ophthalmic divisions of the
      fifth cranial nerves.
      Sympathetic root of the lenticular ganglion.
      Twigs to pituitary body.
Pituitary body, or hypophysis cerebri, 749. (732)
    Anterior lobe.
    Posterior lobe.
    Pituitary gland, pituita.

A pad should be placed in the base of the skull and kept properly moistened with preservative solution, to prevent the structures from drying.

The External Orbital Region.

Examine the eyeball and its surroundings before the skin is removed from the face.

The Eyeball and its Surroundings, 874-5-6-7. (854-5-6-7-8)
    General surface view, Fig. 472.
        Palpebral aperture.
            Outer and inner canthus.
            Lachrymal caruncle. Plica semilunaris.

The Eyeball and its Surroundings :—
    Upper eyelid, tarsus (superior palpebral fold).
    Lower eyelid (inferior palpebral fold).
    Edges of eyelids (a), (b).   (Meibomian follicles.)
    Lachrymal papilla.
    Palpebral conjunctiva, fornix.
    Lachrymal caruncle. Semilunar fold of conjunctiva, 878.   (858)
    Ocular conjunctiva, 876. (856)
    Scleral sulcus.
    Zones of iris; ciliary, pupillary.

An examination of the living eye should be made with the aid of the ophthalmoscope, and the following mentioned structures identified :—

    Optic disc or papilla. Lamina cribrosa.
    Yellow spot. Fovea centralis.

Dissection of the Orbital Region and Face.

Moderately distend the cheeks with tow and stitch the margins of the lips together. Treat the eyelids in the same manner. Continue the incision in the median line of the face to the chin. Beginning at the root of the nose, make a semicircular incision outward along the lower border of the orbit to the outer angle of the eye. Also, make an incision downward in front of the ear, and forward along the lower border of the inferior maxilla or mandible to the median line. Beginning in front of the ear, turn the skin forward to the median line, and remove it. Care must be exercised to remove the skin only, especially over the ala of the nose and the chin, where it is closely adherent to the subjacent structures. The skin of the eyelids should then be removed, by carefully reflecting it to the margin of the lids from above and below. The structures of the external orbital region will be first considered.

Superficial fascia, 452. (447)
Structure of the lids, 1119–20. Fig. 666. (1097) 902. (881)
    Skin. Areolar tissue.
Muscles of the eyelids and eyebrows. Upper tarsal cartilage, 455. (450)

Carefully expose the orbicularis palpebrarum. Preserve the branches of the facial nerve which enter its deep surface, on the outer side.

    Orbicularis muscle, musculus ciliaris Riolini, 902. (881)
    Orbicularis palpebrarum, 455–6; Fig. 309. (450–1)
        Internal tarsal ligament or tendo-oculi.
        External tarsal ligament.
        Orbital portion :—
            Origin and Insertion.
        Palpebral portion :—
            Origin. Insertion.
        Structure. Nerve-supply. Action. Relations.
        Orbicularis, 1120. (1097)

The dissectors of the opposite sides will now work together, and continue the dissection of the external orbital structures on the left side, in the manner indicated. Discontinue further dissection of the right eye at this point, and preserve it for dissection later, with the internal orbital structures.
On the left side, beginning at the outer edge, raise the orbicularis and reflect it inward. Care must be exercised not to injure the palpebral structures underneath. The corrugator supercilii and tensor tarsi will be demonstrated as the orbicularis is raised.

Corrugator supercilii, 457–8; Fig. 310. (452)
    Origin. Insertion. Structure. Nerve-supply. Action. Relations.
Tensor tarsi, 457; Fig. 310. (451)
    Origin. Insertion. Structure. Nerve-supply. Action. Relations.
Central connective tissue, 903. (881) Palpebral ligament, 1120. (1097)
Blood-vessels. Nerves : (a) Sensory; (b) Motor, 904. (883)
Tarsus, superior palpebral muscle, 903. (881–2)

To expose the insertion of the superior palpebral muscle, divide the superior palpebral ligament along the border of the orbit and reflect it downward to the tarsus.

Palpebral conjunctiva, 903. (882)
Lymphatic vessels of the lids, 904. (883)
Glands of the eyelids, 903. (822–3)
    Meibomian glands, lashes, and sebaceous follicles, 1120. (1097)

The Lachrymal Apparatus, 904-5-6 ; Fig. 492.   (883-4-5)
    Lachrymal gland.
Carefully expose the anterior portion of the lachrymal gland in place, raise it slightly, and tease out the minute, thread-like ducts passing from the gland to the conjunctival sac.
Puncta lachrymalia.
Canaliculi lachrymales.
Insert a small probe or a blunt pin in the canaliculi and trace them to the lachrymal sac, opening the canal with the point of the knife.
    Lachrymal sac.
Inner palpebral ligament or tendo oculi, Horner's muscle, 904.   (883)
Outer palpebral ligament, 904.   (883)
Strip the ocular conjunctiva from the surface of the left eye and expose the tendons of the orbital muscles at their insertion. Each tendon can be raised and supported on a small splinter while the insertion is studied. See Fig. 481, P. 890. (869)

Muscles of the orbit, 891 ; Figs. 481, 474.   (870-1)
    Superior and inferior recti.
    External and internal recti.
    Superior and inferior obliques.
    Elevator of the upper lid.

The student will now continue the dissection of the face. The risorius and platysma will first be exposed. The facial branches of the great auricular nerve should be traced.

Risorius, 465 ; Fig. 309.   (459)
    Origin.   Insertion.   Structure.   Nerve-supply.   Action.   Relations.
Platysma myoides, 452-3-4 ; Fig. 309.   (447-8-9)
    (Origin.)   Insertion.   Structure.   Nerve-supply.   Action.   Relations.

Divide the platysma along the lower border of the inferior maxilla or mandible, turn the facial portion forward to its terminal insertion, then remove it.

Facial branches—of the great auricular nerve, 831 ; Fig. 451.   (811)
Parotid fascia, 468.   (462)

Carefully dissect off the fascia, exposing the parotid gland in place, and trace its duct forward to the point where it enters the buccinator. The branches of the facial nerve and the transverse facial artery will be seen emerging from the anterior border of the gland, and in exposing the gland great care must be exercised not to injure these structures. One of the branches of the nerve should be traced back to the main trunk of the facial, cutting away the substance of the gland in order to expose the nerve and other structures passing through it. From the main trunk the radiating branches should be traced forward to their terminal distribution, carefully clearing away the fat and connective-tissue. During the process of exposing the branches of the facial nerve the muscles and vessels of the face will be gradually developed, and, so far as possible, should be exposed in place without injury.

The salivary glands, 982 ; Fig. 559.   (961)
    Parotid gland, 982-3-4.   (961-2-3)
        Superficial area.
        Pterygoid, glenoid, and carotid lobes.   Socia parotidis.
        Duct of the parotid.
        Arteries.   Veins.   Nerves.   Lymphatics.
The facial nerve.
    After its emergence from the stylo-mastoid foramen, the facial nerve runs downward and forward within the substance of the parotid gland, and terminates by dividing into two divisions—an upper or temporo-facial, and a lower or cervico-facial. Six branches are given off from the terminal divisions of the facial nerve :—
        (3), pes anserinus, 809-10 ; Fig. 449.   (790)

Muscles of the mouth, 461 ; Fig. 309.   (455)
    Orbicularis oris, 461.   (456)
    Angular muscles of the mouth, 464 ; Fig. 309.   (458)
        Zygomaticus major, 464.   (458-9)
            Origin.   Insertion.   Structure.   Nerve-supply.   Action.   Relations.
        Depressor anguli oris, 465-6.   (459-60)
            Origin.   Insertion.   Structure.   Nerve-supply.   Action.   Relations.

Muscles of the mouth :—
  Labial group of Muscles, 466. (460)
    Levator labii superioris alaque nasi, 466. (460)
      Origin. Insertion. Structure. Nerve-supply. Action. Relations.
    Levator labii superioris, 466–7. (460–1)
      Origin. Insertion. Structure. Nerve-supply. Action. Relations.
    Zygomaticus minor, 467. (461)
      Origin. Insertion. Structure. Nerve-supply. Action. Relations.

Divide the levator labii superioris transversely, midway between the origin and insertion. Reflect the upper portion to the origin and the lower portion to the insertion, and expose the levator anguli oris. Care must be exercised not to injure the vessels and nerves in reflecting the muscle.

Levator anguli oris, 464–5. (459)
  Origin. Insertion. Structure. Nerve-supply. Action. Relations.
Depressor labii inferioris or quadratus menti, 467. (461)
  Origin. Insertion. Structure. Nerve-supply. Action. Relations.

Expose the buccinator. In cleaning the muscle avoid injury to the nerve branches on its surface. The alveolar origin will be shown; the origin from the pterygo-maxillary ligament—(2)—will be demonstrated at a later period of the dissection.

Buccinator, 462–3–4; Fig. 311. (457–8)
  Origin. Insertion. Structure. Nerve-supply. Action. Relations.

The orbicularis oris should now be fully exposed and its parts demonstrated. To expose the incisive slips, evert the lip and strip off the mucous membrane, when they can be readily shown.

Orbicularis oris, 461–2 ; Fig. 311. (456)
  Attachments to bone :—(1), (2), (3).
  Structure :—
    Fibres :—Transverse set, labial portion.
              Vertical fibres, facial portion.
              Sagittal or antero-posterior fibres, in labial portion.
              Musculi incisivi.
  Nerve-supply. Action. Relations.
Evert the lower lip and draw it downward ; strip off the mucous membrane and expose the levator menti.

Levator menti, 468. (462)
  Origin. Insertion. Structure. Nerve-supply. Action. Relations.
Muscles of the Nose, 459–60–1, 429–30 ; Fig. 516. (453–4–5, 907–8)
  Pyramidalis :—
    Origin. Insertion. Structure. Nerve-supply. Action. Relations.
    Variation.
  Compressor narium :—
    Origin. Insertion. Structure. Nerve-supply. Action. Relations.
  Depressor alæ nasi :—
    Origin. Insertion. Structure. Nerve-supply. Action. Relations.
  Dilator naris anterior :—
    Origin. Insertion. Structure. Nerve-supply. Action. Relations.
  Dilator naris posterior :—
    Origin. Insertion. Structure. Nerve-supply. Action. Relations.
Vessels of the nose, nerves, 930. (909)
  Arteries.
  Veins.
  Nerves.
Facial nerve :—(Motor).
  Temporo-facial division, 810. (790–1)
    Temporal branch.
    Malar branch.
    Infraorbital, infraorbital plexus.
  Cervico-facial division, 810–11. (792)
    Buccal branch.
    Supramandibular branch.
    Inframandibular branch (distribution shown later).

Fifth nerve :—(Sensory branches).
    Ophthalmic division :—
        Infratrochlear nerve, 797. (779)
        Terminal or anterior branch—of nasal nerve, 797. (779)
    Branches from superior maxillary division, 799–800. (780–1)
        Malar branch.
        Labial branches.
        Nasal branches.
        Palpebral branches.
        Infraorbital plexus.
    Branches from the mandibular division :—
        Long buccal nerve, 803. (784–5)
        Mental branch, 804. (786)
Mental branch of mandibular artery, 525. (516–17)
Facial portion of the facial artery, 515–16. (507)
    Branches of the facial artery on the face, 517–18. (509–10)
     . Outer or concave side :—
        Masseteric branches.
        Buccal.
     Inner or convex side :—
        Inferior labial artery.
        Inferior coronary artery.
        Superior coronary artery.
          Arteria septum narium.
        Lateral nasal artery.
        Angular artery.
Angular vein, 650 ; Fig. 385. (636–7)
    Tributaries :—
        Superior lateral nasal veins.
        Palpebral veins.
Facial vein, 650–1 ; Fig. 385. (637–8)
    Tributaries:—
     On the inner side :—
        Inferior lateral nasal vein.
        Superior labial or coronary.
        Inferior labial.
     On the outer side :—
        Inferior palpebral veins.
        Anterior internal maxillary.
        Buccal.
        Anterior parotid.
        Masseteric.
Temporal artery or superficial temporal artery, 521. (513)
    Branches of the temporal artery, 522. (513)
        Parotid branches.
        Articular branches.
        Masseteric.
        Auricular or anterior auricular branches.
        Transverse facial.
Common temporal vein, 652 ; Fig. 385. (639–40)
    Tributaries :—
        Transverse facial vein.　Articular veins.
        Parotid veins.　Masseteric veins.
        Anterior auricular veins.
Temporo-maxillary vein, 653 ; Fig. 385. (640)
Auriculo-temporal nerve, 803. (785)
Superficial lymphatic vessels of the face, 689 ; Fig 398. (674)
Parotid lymphatic glands, 689–90. (674)
Buccal lymphatic glands, 690. (675)

Remove the muscles of the nose and expose the nasal cartilages.

The cartilages, 927-8 ; Figs. 512-13-14. (906-7)
    Superior lateral. Inferior lateral. Cartilage of the septum.
    Also, the sesamoid and accessory quadrate.
        Upper lateral cartilages.
        Lower lateral cartilages.
            Inner plates.
            Outer plates.
        Sesamoid.
        Accessory quadrate.
        Septal cartilage.
        Anterior border.

A pad, which should he kept properly moistened with preservative fluid, should be placed over the orbit and face, and retained in place by a bandage, until the part is wanted for further dissection.

THE EAR, 907. (886)
    External Ear :—Pinna and external auditory meatus.
    Middle Ear, or tympanum, with the Eustachian tube.
    Internal Ear, which includes the osseous labyrinth, within which is placed the
                membranous labyrinth.

The External Ear, 907-8-9-10. (886-7-8-9)
    Pinna, or auricle.
        Helix. Lobule. Concha. Scaphoid fossa.
        Antihelix. Antitragus. Triangular fossa. Tragus. Antitragus.

The skin should be removed from the surface of the entire auricle, exposing the structures underneath.

        Cartilaginous framework. Lobule. Fissures of Santorini.
        Ligaments :—
            Anterior ligament.
            Posterior.
        Muscles :—Extrinsic muscles (attollens, attrahens, and retrahens aurem).
            Intrinsic muscles :—Helicis major. Helicis minor.
                        Tragicus. Antitragicus. Transversus auris.
                        Obliquus auris.
        Vessels :—
            Arteries :—Posterior auricular. Anterior auricular.
            Veins.   Also, Veins of the ear, 663. (650)
        Nerves :—Great auricular. Posterior auricular of the facial.
            Auricular branch of the pneumogastric.
            Auriculo temporal. Small occipital.
        (External Auditory Meatus.)

The pinna should be left in place. The external auditory canal will be more fully considered at a later period, in connection with the middle ear.

THE NECK.

Superficial Structures.
Landmarks in the middle line, 1124-5-6. (1101-2-3)

Continue the median incision to the sternum and make a transverse incision along the anterior border of the clavicle to the outer end. Turn the skin outward and backward from the median line, exposing the superficial fascia. The platysma should then be exposed.

Superficial fascia, 452. (447)
Platysma myoides, 452-3-4. (447-8-9)
    (Origin.) Insertion. Structure. Nerve-supply. Action. Relations. Variations.

Beginning at the mandible, carefully reflect the platysma downward to the clavicle and remove the cervical portion, dividing it about half-an-inch above the clavicle. In reflecting the platysma great care must be exercised not to remove or injure the nerves or veins immediately beneath it.

External jugular vein, 654. (641)
    Chief variations of the external jugular vein.

External jugular vein :—
>Tributaries and communications, Fig. 385.
>>Posterior external jugular.
>>Transverse cervical and suprascapular.
>>Sometimes—the anterior jugular.
>Posterior external jugular vein.

Vertical set of superficial cervical glands of the neck, or superficial cervical chain, 691. (675-6)

Anterior jugular vein, 655. (642)

Cervical plexus :—
>Upper four cervical nerves, 829.   (809-10)
>Superficial Branches, 829-30-1.   (810-11-12)
>>Ascending branches :—
>>>Lesser occipital nerve.
>>>>Auricular branch.
>>>>Mastoid branch.
>>>>Occipital branch.
>>>Great auricular nerve :—
>>>>Mastoid branch.
>>>>Auricular branches.
>>>>· Facial branches.
>>Transverse branch :—
>>>Superficial cervical nerve.
>>Descending branches :—
>>>Suprasternal twigs.
>>>Supraclavicular nerves.
>>>Supra-acromial branches.

Inframandibular branch—of the facial nerve, 811 ; Fig. 449. (792)

Read :—The cervical fascia.   Prevertebral fascia, 472-3.   (466)
>Deep cervical fascia, 1130-1-2-3-4.   (1107-8-9-10-11)
>>Arrangement above the hyoid bone :—
>>>Superficial.
>>>Deeper layer.
>>Below the hyoid bone :—
>>>Four layers :—Superficial or subcutaneous.
>>>>Sterno-clavicular.
>>>>Tracheal.
>>>>Prevertebral.
>>Traced vertically :—
>>>At the level of the top of the sternum :—
>>>>Superficial.
>>>>Sterno-clavicular.
>>>>Tracheal.
>>>>Prevertebral.
>>>At the level of the clavicle :—
>>>>Subcutaneous.
>>>>Sterno-clavicular.
>>Uses and important points :—
>>>(A) Forms certain definitely enclosed spaces.
>>>(B) Sheaths of canals.
>>>(C) Helps to resist atmospheric pressure.
>>>(D) Action on the pericardium, prevents pressure of the lungs on the heart.

The relations of the deep fascia should be remembered and recalled as the structures enclosed by it are successively exposed during the process of the dissection.

Reflect the fascia and expose the sterno-cleido mastoid in place. The external jugular vein should be preserved.

Sterno-cleido-mastoid, 473-4-5. (467-8) Also, 1126-7. (1103)
 Origin; sternal head, clavicular head. Insertion. Structure. Nerve-supply.
 Action. Relations. Variations.

The sterno-cleido-mastoid divides the side of the neck into an anterior and a posterior triangular space. The boundaries of the posterior space are formed by the clavicle below, the sterno-cleido-mastoid in front, the trapezius behind, and the occiput above. The posterior belly of the omo-hyoid crosses the space about an inch above the clavicle, and divides it into an upper or occipital, and a lower or subclavian triangle. The floor of the space is formed from above downward by the splenius, levator anguli scapulæ, scalenus medius, and posticus.

Posterior Triangle, 1129. (1106)

Carefully remove the fascia and dissect out the connective tissue and fat, exposing the structures in the posterior triangular space.

Omo-hyoid, 476 ; Fig. 314. (469)
 (Origin. Insertion. Structure. Relations.)
Spinal accessory nerve, 1130. (1106-7)
Branches to the trapezius, 831. (812)
Nerves to the levator anguli scapulæ, 831. (812)
Third portion of the subclavian artery, 538-9. (529-30)
 Relations :—In front.
    Behind.
    Below.
    Above.

The outer portion of the transverse cervical and suprascapular arteries will be exposed, the origin of the vessels will be shown later.

Transverse cervical or transversalis colli artery, 546-7. (537)
Suprascapular or transversalis humeri, 546. (536)
 Branches of the suprascapular :—
    (1), (2), (3), (4).
Veins :—
  External jugular vein, 1130. (1106)
  Suprascapular veins, 654. (641)
  Transverse cervical veins, 654. (642)
  Subclavian vein, 682-3. (668)
   Tributaries.

Anterior Triangular Space.

The muscles forming the boundaries of the smaller triangles should first be exposed. Remove the fascia forming the sheaths of the muscles and dissect out the fat and connective tissue. The small nerves, the arterial branches, and the venous tributaries must all be carefully preserved. Beginning at the symphysis, expose the anterior belly of the digastric ; then the posterior belly anterior to the border of the sterno-mastoid, removing the portion of the parotid gland superficial to it. The process of fascia binding the tendon to the hyoid bone should be preserved. In exposing the omo-hyoid avoid injury to the branches of the ansa hypoglossi which pass to it.

Digastric, 478-9 ; Fig. 314. (471-2)
 Origin : anterior belly (posterior belly). Insertion. Structure.
 Nerve-supply. Action. Relations. .Variations.
Stylo-hyoid, 479 ; Fig. 314. (472)
 Origin. Insertion. Structure. Nerve-supply. Action. Relations. Variations.
Omo-hyoid, 476 ; Fig. 314. (469-70)
 (Origin.) Insertion. Structure. Nerve-supply. Action. Relations. Variations.
Anterior triangle, 1127-8-9. (1104-5-6)
 Subdivided into three triangles :—
  Submaxillary or supra-hyoid triangle.
  Superior carotid triangle.
  Inferior carotid or tracheal triangle.

Before continuing the dissection the student requires some preliminary knowledge of the structures in the anterior triangle, their position, relations, etc. The following general description (modified by Cunningham) should be carefully studied.

General Description.

Inferior Carotid or Tracheal Triangle.—As this triangle is gradually opened up the following structures will come into view :—

1. The sterno-hyoid and sterno-thyroid muscles.
2. The branches from the ansa hypoglossi to these muscles.
3. The external laryngeal nerve.
4. The superior thyroid artery.
5. The greater part of the larynx, the thyroid body, and the trachea.
6. The œsophagus on the left side.
7. The recurrent laryngeal nerve.

When the fascia is removed from this part of the anterior triangle, the only structures which are seen within its limits are the sterno-hyoid and sterno-thyroid muscles. As the dissection is proceeded with, however, the large nerve of supply for these muscles, which comes from the ansa hypoglossi, will be found lying near the outer border of the sterno-thyroid and breaking up into numerous twigs. Toward the upper part of the space the superior thyroid artery will be noticed passing downward under cover of the omo-hyoid, sterno-thyroid, and sterno-hyoid muscles to reach the thyroid body. At a slightly higher level than this artery, the external laryngeal nerve runs forward to end in the crico-thyroid muscle. Under cover of the sterno-hyoid and sterno-thyroid muscles will be observed the larynx, the isthmus and a considerable part of the lateral lobe of the thyroid body, and the trachea. The recurrent laryngeal nerve lies deeply. It will be found in the interval between the gullet and the trachea. As the œsophagus inclines to the left behind the trachea, it follows that it is only seen, when in its natural position, in the left inferior carotid or tracheal triangle.

Superior Carotid Triangle.—During the dissection of the superior carotid triangle the following parts are displayed:—

Arteries.
1. Common carotid dividing into external and internal carotid arteries.
2. Superior thyroid. ⎫
3. Lingual. ⎪
4. Facial. ⎬ Branches of the external carotid.
5. Occipital. ⎪
6. Ascending pharyngeal. ⎭
7. Hyoid. ⎫
8. Sterno-mastoid. ⎬ Branches of the superior thyroid.
9. Superior laryngeal. ⎭
10. Sterno-mastoid branch of the occipital.

Veins.
1. Internal jugular.
2. Facial. ⎫
3. Anterior temporo-maxillary. ⎬ Tributaries of the internal jugular.
4. Lingual. ⎪
5. Superior thyroid. ⎭

Nerves.
1. Hypoglossal.   Crossing the space in a transverse direction.
2. Descendens hypoglossi. ⎫
3. Nerve to thyro-hyoid. ⎪
4. Internal laryngeal. ⎬ Crossing the space obliquely.
5. External laryngeal. ⎪
6. Spinal accessory. ⎭
7. Vagus. ⎫ Descending vertically.
8. Sympathetic. ⎭

Intercarotic body.
Portion of the larynx and pharynx.
Greater cornu of the hyoid bone.
Lymphatic vessels and glands.

This subdivision of the anterior triangle, when opened up, contains portions of each of the three carotid arteries. It is well, however, that the dissector should clearly understand that it is only after the parts are relaxed by dissection that these vessels come to lie within the space. When the fascia and platysma are in position they are completely overlapped by the sterno-mastoid muscle. The common carotid artery is enveloped along with the internal jugular vein and the vagus nerve in a common sheath of fascia. The carotid sheath should be slit open, care being taken of the descendens hypoglossi nerve, which also descends obliquely within it. The intimate connection which this sheath presents with the prevertebral layer of fascia can now be made out. The common carotid usually bifurcates opposite to the upper border of the thyroid cartilage. Observe that the internal carotid at first lies *behind, and to the outer side* of the external carotid. Upon the coats of these vessels numerous sympathetic twigs ramify, and, at the point of bifurcation of the common carotid, a small oval body, termed the *intercarotic body*, will be found closely applied to the deep surface of the vessel. The branches of the external carotid which take origin in the area of this triangle run for only a very short part of their course within it. Three branches will be noticed springing from the anterior aspect of the external carotid. Named from below upward these are: (1) the *superior thyroid*, which, after having given off its *hyoid, superior laryngeal*, and *sterno-mastoid branches*, disappears under cover of the omo-hyoid muscle, and enters the inferior carotid triangle; (2) the *lingual*, which leaves the space by passing under cover of the digastric and stylo-hyoid muscles; and (3) the *facial*, which ascends under the same muscles to gain the digastric triangle. The *occipital artery* commonly springs from the posterior aspect of the external carotid close to the lower border of the digastric, and soon quits the space by running upward and backward under cover of the sterno-mastoid muscle. Its *sterno-mastoid branch* comes off as it leaves the triangle, and accompanies the spinal accessory nerve. The *ascending pharyngeal artery* will be found by separating the external and internal arteries from each other, and dissecting between them. It springs from the posterior aspect of the former about half-an-inch above its origin, and then takes a vertical course upward on the prevertebral muscles.

The *internal jugular vein* lies close to the outer side of the common carotid artery, and is included within the same fascial sheath. Several tributaries join it as it passes through this space. The most conspicuous of these is the *common facial vein*, which is formed by the union of the facial vein and the anterior division of the temporo-maxillary vein.

The nerves which are brought into view as the carotid triangle is gradually opened up are very numerous, but they can be classified according to the direction which they take through the space. One large nerve, the *hypoglossal*, takes a more or less *transverse* course across the upper part of the space. It forms a loop across this part of the neck immediately below the lower margin of the digastric muscle. Two descend *vertically*,—viz., the vagus and the gangliated cord of the sympathetic. The *vagus* lies in the posterior part of the carotid sheath between the common carotid artery and the internal jugular vein. The *sympathetic cord* is imbedded in the posterior wall of the carotid sheath. The remaining five nerves traverse the triangle obliquely. Four run from above downward and forward, viz., the descendens hypoglossi, the thyro-hyoid, the internal and external laryngeal nerves; and one—the spinal accessory—is directed from above downward and backward.

The *descendens hypoglossi* springs from the hypoglossal nerve as it hooks round the occipital artery and descends within the carotid sheath. It is joined at the upper border of the omo-hyoid by one or two branches from the cervical plexus, and in this manner the *ansa hypoglossi* is formed. The *thyro hyoid* is a minute nerve which arises from the hypoglossal nerve before it disappears under cover of the posterior belly of the digastric. The *internal laryngeal nerve* descends behind the carotid vessels. It will be readily found, with the corresponding artery, in the interval between the hyoid bone and upper border of the thyroid cartilage. It enters the larynx by piercing the thyro-hyoid membrane. The *external laryngeal nerve*, a branch of the preceding, is a much smaller twig. It leaves the space by passing under cover of the depressor muscles of the larynx and hyoid bone. The *spinal accessory nerve* is placed high up in the interval between the digastric and the sterno-mastoid muscles. It soon disappears by sinking into the substance of the sterno-mastoid.

The Submaxillary or Supra-hyoid Triangle may be divided very conveniently into an anterior and posterior part by the stylo-maxillary ligament and a line drawn downward from it. The portion in front has a distinct floor, composed, in great part, by the mylo-hyoid muscle, and behind this by a portion of the hyoglossus muscle.

The parts exposed during the dissection of the digastric space may be classified according to the subdivision in which they lie :—

ANTERIOR PART.	POSTERIOR PART.
1. Submaxillary gland.	1. Portion of the parotid gland.
2. Facial artery and vein.	2. External carotid artery.
3. Branches from facial artery in this part of its course. *a.* Ascending palatine. *b.* Tonsillitic. *c.* Submaxillary. *d.* Submental.	3. Posterior auricular artery.
4. Hypoglossal nerve.	
5. Mylo-hyoid nerve.	
6. Mylo-hyoid artery.	
7. Lymphatic glands.	

The most conspicuous object in the fore part of the digastric triangle is the *submaxillary gland*. The *facial artery* passes upward and forward in the midst of this gland, whilst the *facial vein* lies superficial to it. But whilst the facial artery runs through the gland, the dissector should note that it can be separated from it without any laceration of the gland lobules. The gland is, as it were, wrapped round it, so that, although at first sight the artery seems to pierce it, it is in reality merely contained in a deep furrow in its substance. Before entering the gland, the facial artery gives off its *tonsillitic* and *ascending palatine branches*, whilst its *submental* and *submaxillary branches* arise within the gland. The submental artery runs forward toward the chin. The *mylo-hyoid nerve* and *artery* will be seen passing forward and downward upon the mylo-hyoid muscle under cover of the submaxillary gland. The twigs of the nerve to the muscle of the same name and to the anterior belly of the digastric should be followed out. Only a very small part of the *hypoglossal nerve* is seen in this space. It lies upon the hyoglossus muscle immediately above the hyoid bone, and disappears under cover of the posterior border of the mylo-hyoid muscle. Numerous small *lymphatic glands* lie under shelter of the base of the lower jaw.

The *external carotid artery* enters the posterior part of the digastric triangle. Here it lies under cover of the lower part of the parotid gland, and gives off its *posterior auricular branch*, which passes upward and backward along the upper border of the posterior belly of the digastric muscle.

Middle Line of the Neck.—Before the parts are further disturbed the dissector should examine the structures which occupy the middle line of the neck—a region, more especially in its lower part, of the highest importance and interest to the surgeon. The middle line of the neck may be divided by means of the hyoid bone into an upper supra-hyoid and a lower infra-hyoid portion.

In the *supra-hyoid part* are found structures which are concerned in the construction of the floor of the mouth. The student has already noticed that the fatty superficial fascia is more fully developed here than elsewhere in the neck, and that the anterior margins of the two platysma muscles meet in the mesial plane about half-an-inch or so below the chin. Above this point their fibres decussate. In the present condition of parts the two anterior bellies of the digastric muscles are observed attached to the mandible on either side of the symphysis. From this they descend toward the hyoid bone, and diverge slightly from each other so as to leave a narrow triangular space between them. The floor of this space is formed by anterior portions of the two mylo-hyoid muscles, whilst bisecting the triangle in the mesial plane in the fibrous raphe, into which these muscles are inserted. Not unfrequently the inner margins of the digastric muscles send decussating fibres across the interval between them.

The *infra-hyoid part* of the middle line of the neck extends from the hyoid bone to the upper margin of the manubrium sterni. Below the hyoid bone is the thyro-hyoid membrane, succeeded by the thyroid cartilage with its prominent *pomum Adami*. Then comes the tense crico-thyroid membrane and the cricoid cartilage. Except along a narrow interval on either side of the mesial plane, these structures are covered with two muscular strata, viz., a superficial layer formed by the sterno-hyoid and omo-hyoid, which lie on the same plane, and a deeper layer formed by the sterno-thyroid and its continuation upward, the thyro-hyoid. In addition to these an elongated process of the thyroid gland not unfrequently stretches upward (usually upon the left side), under cover of the sterno-hyoid. If this be present, it will in all probability be observed to be attached to the hyoid bone by a narrow muscular band, termed the *levator glandulæ thyroideæ*. The divergent inner margins of the two small crico-thyroid muscles, as they extend upward and outward upon the cricoid cartilage toward the lower margin of the thyroid cartilage, will also be seen. Upon the crico-thyroid membrane, as it appears between these muscles, the small crico-thyroid artery runs transversely inward.

Below the cricoid cartilage the dissector comes upon the trachea, which extends downward through the remaining portion of the middle of the neck. As the tube descends it gradually recedes from the surface, so that at the upper margin of the sternum it lies very deeply. The length of this portion of the trachea varies with the position of the head. When the chin is raised, and the head thrown as far back as possible, about two inches and three-quarters of the tube will be found between the cricoid cartilage and the sternum ; when, on the other hand, the head is held in its usual attitude, the length of the cervical part of the trachea is diminished by fully three-quarters of an inch. These measurements must be regarded as merely express-ing the average condition. They vary considerably in different individuals, and are much influenced by differences in the length of the neck.

The dissector must study carefully the parts which lie superficial to the trachea in this portion of its course. In the first place, examine the structures which are in immediate contact with it. These are (1) the isthmus of the thyroid body; (2) the inferior thyroid veins; (3) at the root of the neck, the innominate artery and the left innominate vein ; (4) the thymus body in young children ; and (5) the occa-sional thyroidea ima artery. The *isthmus of the thyroid* is a thin band of thyroid substance which crosses the mesial plane upon the anterior aspect of the trachea. As a general rule it covers the second, third, and fourth tracheal rings, so that only one ring is left exposed between its upper margin and the cricoid cartilage. Almost in-variably a branch of the superior thyroid artery runs along the upper margin of the isthmus, whilst upon its anterior surface is placed a plexiform arrangement of small veins. The *inferior thyroid veins* are two in number and of large size. They are formed by several tributaries which issue from the lateral lobes, and proceed down-ward upon the front of the trachea, one upon either side of the mesial plane. They are separated from each other by a narrow interval, and immediately below the isth-mus they are connected by a plexus of small veins which lies in front of the trachea. An additional *median vein*, taking origin from the isthmus, may also exist. Close to the sternum the *innominate artery* will be observed lying upon the trachea, and slightly below the level of the upper border of the bone the *left innominate vein* crosses it. The *thyroidea ima* is an occasional branch of the innominate artery. When present it passes vertically upward in front of the trachea to the isthmus of the thyroid gland.

The parts which separate the trachea, with the structures in immediate relation to its anterior aspect, from the surface, should now be studied. The two anterior jugular veins as they run downward in the superficial fascia, one upon either side of the mesial plane, have been already noticed ; also, the two layers of the deep cervical fascia close to the upper margin of the sternum, and in the interval between these the cross-branch connecting the two anterior jugular veins. Behind the fascial envelope of the neck come the two muscular strata formed by the sterno-hyoid and the sterno-thyroid muscles. The inner margins of the sterno-hyoid muscles are almost contiguous above, and held together by the fascial sheaths which enclose them ; below, however, they diverge slightly from each other, so as to expose, close to the sternum, the inner margins of the sterno-thyroid muscles. The sterno-thyroid muscles, in contact with each other below, gradually separate from each other as they

3

ascend. A narrow, lozenge-shaped space is thus left between the inner borders of these muscles. Over this area the trachea is not covered by any muscular structure.

SURGICAL ANATOMY. (From *Cunningham.*)

The principal operations which are performed in the middle line of the neck are those of laryngotomy and tracheotomy.

In laryngotomy, an opening is made into the larynx. This can most readily be done in the interval between the thyroid and cricoid cartilages. A vertical mesial incision through the integument is made over this interval. The crico-thyroid membrane is thus exposed, and is divided *transversely* close to the upper margin of the cricoid cartilage. It is a very simple proceeding, and one which is attended with little or no danger if ordinary care be taken. The crico-thyroid membrane is divided transversely, and in its lower part, for two reasons, viz., (1) to avoid injury to the crico-thyroid artery, which, although, as a general rule, of small size and of no surgical importance, is sometimes large enough to give rise to awkward results if wounded ; and (2) to place the opening as low down as possible.

Tracheotomy is a more serious operation. The opening into the trachea may be made above or below the isthmus of the thyroid body. The high operation is very properly preferred by the surgeon. Its advantages are very apparent : here the trachea lies near the surface, and no veins of any importance are met with. The only drawback consists in the small portion of trachea which intervenes between the isthmus and the cricoid cartilage. Still, this can be increased by pushing down the isthmus, which, within certain limits, can be easily dislodged in a downward direction. Many surgeons, indeed, consider that the wounding of the isthmus is a matter of comparatively slight importance. The fact, however, that a large branch of the superior thyroid artery is generally found in relation to its upper border should make the operator hesitate before having recourse, in all cases, to this expedient for gaining additional space. In the child it is frequently necessary to combine the high operation of tracheotomy with that of laryngotomy—viz., by cutting through the cricoid cartilage.

The low operation is a formidable undertaking. It is true that there is a greater length of tube to be operated upon; but this is situated very deeply, and the surgeon encounters many difficulties before it is reached. If the dissector reflect upon the structures which intervene between this part of the trachea and the surface, he will fully realize this; and he must bear in mind that these difficulties are greatly intensified in the living subject by the engorged state of the veins and the convulsive movements of the windpipe as the patient struggles for breath. In the child, the thymus body interposes an additional obstacle; and this, combined with the more limited space, the small calibre and great mobility of the trachea, render the operation, in such cases, a very serious responsibility. In the low operation, the trachea must be opened in an upward direction, so as to avoid injury to the innominate artery and left innominate vein, which are placed in front of it at the upper margin of the sternum.

Dissection.

" Dissection.—The numerous and diverse structures contained within the anterior triangle must now be displayed. The dissection should be carried out over the entire area at once, and the structures found in one subdivision followed upward or downward, as the case may be, into the other subdivisions of the space. It is a common fault with dissectors to fail to open up the digastric triangle until the two lower triangles have been fully dissected. Two small nerves are especially liable to injury, and therefore should be secured as early as possible. They are the thyro-hyoid branch of the hypoglossal nerve, and the external laryngeal nerve. The hypoglossal nerve, which will be found crossing the carotid triangle at the lower border of the posterior belly of the digastric, should be traced forward ; as it approaches the hyoid bone, its minute thyro-hyoid branch will be discovered, leaving its lower border at an acute angle, and proceeding downward and forward to reach the thyro-hyoid muscle. The external laryngeal nerve is a long, slender branch which occupies a deeper plane. To expose it the carotid vessels should be pulled outward from the larynx, and the loose tissue in the interval thus opened up divided carefully in an oblique direction and along a line connecting the cricoid cartilage with the bifurcation of the common carotid artery. The nerve will be found as it passes downward and forward to disappear under cover of the depressor muscles of the larynx. It will be traced to its ultimate distribution at a subsequent stage of the dissection." (Cunningham.)

Hypoglossal nerve, 820–1. (801–2)
    Descendens hypoglossi :—
        Ansa hypoglossi.
Sterno-hyoid, 475–6. (468–9)
    Origin. Insertion. Structure. Nerve-supply. Action. Relations. Variations.

Divide the sterno-hyoid transversely about an inch above the sternum, raise the upper portion as far as the insertion, reflect the lower portion to the origin. Expose the sterno-thyroid and thyro-hyoid muscles, then replace the sterno-hyoid.

Sterno-thyroid, 477. (470)
    Origin. Insertion. Structure. Nerve-supply. Action. Relations. Variations.
Thyro-hyoid, 477. (470–1)
    Origin. Insertion. Structure. Nerve-supply. Action. Relations. Variations.

Divide the sternal origin of the sterno-mastoid. Remove the tissues covering the sterno-clavicular joint and expose the ligaments. The clavicular attachment of the sterno-mastoid should be preserved until the dissection of the subclavian and carotid arteries is completed, and their relations studied.

Sterno-clavicular articulation, 227-8-9-30.   (233-4-5-6)
  Class.—*Diarthrosis.*   Subdivision.—*Arthrodia.*
   Capsular ligament:—
    Posterior sterno-clavicular ligament.
    Anterior sterno-clavicular ligament.
     Below.   Superior portion.
    Interclavicular ligament.
    Rhomboid or costo-clavicular ligament.

Remove the anterior ligament and expose the interarticular fibro-cartilage within the joint.

    Interarticular fibro-cartilage.
     Synovial membrane.
  Arterial supply.
  Nerve-supply.
  Movements.
Behind the sterno-clavicular joint, 1127.   (1103-4)

Carefully divide the remaining ligaments of the sterno-clavicular articulation, and draw the sternal end of the clavicle slightly forward; with the saw divide the clavicle by an oblique incision extending from the outer margin of the sterno-cleido-mastoid to the lower portion of the sternal extremity of the bone. Raise the fragment of clavicle with the attached sterno-cleido-mastoid, and, carefully separating the muscle from the structures underneath, turn it upward to its insertion. In raising the muscle, note the vessels and nerves entering its deep surface. The sterno-mastoid must not be detached at its insertion until the dissection of the subclavian and carotid vessels is completed. When these vessels are exposed, replace the sterno-mastoid in position while studying their relations. As the jugular and subclavian veins lie in a plane superficial to the arteries, they may be first exposed, and then pushed aside while the more important vessels, the arterial trunks, are being displayed. Beginning above, trace the terminal or exposed portion of the tributaries to the internal jugular, and follow the internal jugular from the posterior belly of the digastric to its termination; the upper portion of the vessel will be exposed at a later period of the dissection.

Facial vein, common facial vein, 650.   (637-8)
  Submental vein, 651.   (638)
  Inferior or descending palatine vein, 651.
  Submaxillary or glandular veins, 651.
  Communicating branch, or anterior division of the temporo-maxillary vein, 651.
  Chief variations in the facial vein.

Lingual vein, 665-6.   (652)
Superior thyroid vein, 666.   (652)
Middle thyroid vein, 666.   (652)
Internal jugular vein, 665.   (651-2)
  Tributaries.

Expose the scaleni muscles and the subclavian vessels. Care must be exercised not to injure the small branches of the cervical or the brachial plexus, or the branches of the subclavian artery. A portion of the pleura of the lung will also be exposed, but must not be injured. The termination of the thoracic duct will be found at the junction of the subclavian and internal jugular veins.

Scalenus anticus, 484; Fig. 316.   (476-7)
  Origin. Insertion. Structure. Nerve-supply. Action. Relations.
Scalenus medius, 484; Fig. 316.
  Origin. Insertion. Structure. Nerve-supply. Action. Relations.
Scalenus posticus, 484-5.   (477)
  Origin. Insertion. Structure. Nerve-supply. Action. Relations.
 Variations of the scaleni.

Subclavian vein, 682-3.   (668-9)
  Tributaries.
  Chief variations in the subclavian vein.

Cervical portion of the thoracic duct, 699.   (683-4)

Right lymphatic duct, 688, 692–3. (673, 677)
Subclavian artery, 536. (527)
    First portion of the right subclavian artery, 537. (528)
        Relations:—In front.
                Behind.
                Below.
    Branches.

The upper or cervical part of the first portion of the left subclavian artery will be exposed. The origin from the aorta will be shown at another time. The relation to the sterno-mastoid should be demonstrated by replacing the muscle in position. The right subclavian can be exposed from its origin.

Left subclavian artery, 536–7. (527–8)
        Relations:—In front.
                Behind.
                Right side.
                Left side.
Second portion of the subclavian artery, 538. (528)
        Relations:—In front.
                Behind.
                Above.
                Below.
    Branch.
Third portion of the subclavian artery, 538–9. (529–30)
        Relations:—In front.
                Behind.
                Below.
                Above.

Chief Variations in the Subclavian Artery.

Expose the branches of the subclavian artery, and trace them to the point where they pass into or under other structures; the terminal portion of such branches will be demonstrated later in the process of the dissection.

Branches of the subclavian artery, 540. (530)
    First portion:—Vertebral.
                Thyroid axis.
                Internal mammary.
    Second portion:—Superior intercostal.
    Third portion.

Vertebral artery, 540. (530–1)
    The first, or cervical portion:—
                In front.
                Behind.
                Inner side.
                Outer side.

Chief Variations of the Vertebral Artery, 541. (*532*)

Vertebral vein, 666. (653)
    Tributaries:—Anterior vertebral vein.
                Deep cervical vein, 652. (639)
Thyroid axis, 545. (535)
    Inferior thyroid.
    Suprascapular.
    Transverse cervical.
Inferior thyroid, 545–6. (536)
    Branches:—Muscular branches.
             Ascending cervical.
                Muscular.   Spinal.   Phrenic.
            Œsophageal branches.

Inferior Thyroid Branches:—
 Tracheal branches.
 Inferior laryngeal branch.
Inferior thyroid veins, 666. (652)
 Right vein.
 Left vein.

Suprascapular or transversalis humeri, 546. (536)
Suprascapular veins, 654. (641) .
Transverse cervical or transversalis colli, 546-7.
Transverse cervical veins, 654. (642)
Internal mammary artery, 548. (538)
 Cervical portion.
Superior intercostal artery, 550-1-2. (540-1-2)
 Branches:—Deep cervical branch :—
 (Muscular. Anastomotic. Vertebral or spinal).
 First intercostal branch.
 Arteria aberrans.

The superficial branches of the cervical plexus have already been displayed; some of the deep branches may now be exposed ; others will be shown later, as the deeper structures are dissected.

Cervical nerves, 828-9. (809)
 Upper four nerves:—
 First nerve. Second nerve. Third and fourth nerves.
Cervical plexus, 829. (809-10)
 Upper four cervical nerves.

The following table (from Cunningham) indicates the origin, branches, and distribution of the cervical nerves :—

Superficial or Cutaneous.	Ascending.	Small occipital.	From 2d.
		Great auricular.	From 2d and 3d.
	Transverse.	Superficial cervical.	
	Descending.	Supraclavicular.	
		Suprasternal.	From 3d and 4th.
		Supra-acromial.	

Deep.	Muscular.	1. To rectus capitis anticus major.	From 1st and 2d.
		2. To rectus capitis anticus minor.	
		3. To rectus capitis lateralis.	
		4. To sterno-mastoid.	From 2d.
		5. Communicantes hypoglossi.	From 2d and 3d.
		6. To levator anguli scapulæ.	From 3d and 4th.
		7. To scalenus medius.	
		8. To trapezius.	
		9. Phrenic, to diaphragm.	From 4th and 5th.
	Communicating.	1. To hypoglossal.	From 1st.
		2. To vagus.	
		3. To sympathetic.	From 1st, 2d, 3d, and 4th.
		4. To spinal accessory.	From 2d, 3d, and 4th.

Deep branches, 831-2. (812-13)
 Nerve to sterno-mastoid.

Cervical Nerves, Deep Branches:—
        Nerves to the scalenus medius.
        Nerves to the levator angulæ scapulæ.
        Branches to the trapezius.
        (Internal branches):—
            Communicating branches.
            Muscular branches :—
                Branch to the rectus capitis lateralis.
                Nerve to the rectus capitis anticus minor.
                Nerve to the rectus capitis anticus major.
                Nerve to the longus colli.
             † { Communicantes hypoglossi, ansa hypoglossi.
               { Phrenic nerve.

                † Communicans hypoglossi and phrenic nerves should be traced; the other
            branches can be shown in the dissection of the deeper structures.

The common carotid artery and its divisions should now be exposed. The origin and thoracic portion of the left common carotid will be shown at another time. The origin of the branches of the external carotid should be exposed, and the branches should be traced to the point where they pass into or under other structures. The terminal distribution of such branches will be developed later in the dissection. The pneumogastric nerve and its branches will be considered in the deep dissection of the neck; the branches must be carefully preserved.

Common carotid arteries, 504–5. (496–7)
        External carotid.
        Internal carotid.
(Thoracic portion of the left common carotid artery), 505. (497)
Common carotid artery in the neck, 506–7–8–9. (498–9–500–1)

      When the artery and its divisions have been fully exposed, the sterno-mastoid should be
      replaced in position and the relations carefully studied.

        Relations:—In front.
             Behind.
             Internally :—
                Ganglion intercaroticum, or carotid gland.
             Externally.
        Branches.

        Variations of the Common Carotid Arteries.
        The collateral circulation.

External carotid artery, 509–10. (501–2)
       Relations:—In front.
           Behind.
           Internally.
           Externally.

    Chief Variations of the External Carotid Artery.

        Branches of the external carotid, 510. (502)
            Superior thyroid artery, 512–13. (503–4–5)
                Branches of the superior thyroid artery :—
                  Hyoid—or infra-hyoid branch.
                  Sterno-mastoid—or middle mastoid artery.
                  Superior laryngeal artery.
                  Crico-thyroid—or inferior laryngeal branch.
           Lingual artery, 513–14. (505–6)
             First or oblique portion :—
                Hyoid, or supra-hyoid.
           Facial artery, 515–16–17. (507–8–9)
             Cervical portion :—
                Branches of the facial artery in the neck :—
                  (Inferior or ascending palatine.)
                (Tonsillar branch.)

Branches of the Facial Artery in the Neck :—
 Glandular or submaxillary branches.
 Muscular branches.
 Submental branch :—
  Branches :—Muscular. Perforating.
       Cutaneous. Mental.
Occipital artery, 518. (510-11)
 First part of its course.
 Sterno-mastoid branch.
 Auricular branch.
Posterior auricular artery, 520-1. (512-13)
 Branches of the posterior auricular artery :—
  Parotid branches.
  Muscular branches.
  Stylo-mastoid branch :—
   (Meatal. Mastoid. Stapedic.)
   (Tympanic. Vestibular. Terminal.)
Ascending pharyngeal artery, 510. (502)
(Terminal divisions of the external carotid :—)
 Temporal artery, 521. (513)
 Internal maxillary artery, 522. (514)
Deep lymphatic vessels of the head and neck, 691. (676)
Deep lymphatic glands of the neck, 692. (677)
 Upper set.
 Lower deep cervical glands.

Divide the sterno-thyroid transversely about an inch above the sternum; raise the upper portion to the insertion, exposing the thyroid body or gland.

The thyroid body or gland, 952-3-4-5. (931-3-4-5)
 Lateral lobes ; isthmus.
 Relations.
 Middle or pyramidal process, thyro-glossal duct.
 Structure.
 Vessels :—
  Arteries.
  Veins.
 Nerves.
 Lymphatics of the thyroid body, 692. (677)

The cervical portions of the trachea and œsophagus may be exposed; in the interval between them will be found the cervical portion of the recurrent laryngeal nerve. The cervical portion of the recurrent laryngeal nerve should be exposed ; the origin and the terminal distribution will be seen later.

The trachea, or air tube, 950-1. (929)
 Relations :—Cervical portion.
Inferior or recurrent laryngeal nerve, 818-19. (799-800)
 Of the right side.
 Left side.
Œsophagus, 987-8. (966)
 Relations in the neck :—
  In front.
  Behind.
  On its right side.
  On the left side.
Lymphatics of the upper part of the œsophagus and trachea, 692. (677)

Temporal and Pterygo-maxillary regions.

Divide the nerves and vessels crossing the masseter near its posterior border and throw them forward. Remove the fascia and expose the surface of the muscle.

Muscles of mastication, 468. (462)
> Temporal fascia.
> Masseteric fascia.

Masseter, 469. (462–3)
> Origin. Insertion. Structure. Nerve-supply. Action. Relations.

Divide the temporal fascia along the upper border of the zygoma, reflect it upward and remove it, exposing the upper portion of the temporal muscle. Saw through the zygomatic arch at the anterior and posterior borders of the masseter and turn the muscle downward to its insertion; in doing this, note the vessels and nerves entering its deep surface, then divide them close to the muscle. The insertion of the temporal will then be exposed.

Temporal muscle, 469–70. (463–4)
> Origin. Insertion. Structure. Nerve-supply. Action. Relations.

With the saw, divide the coronoid process by an oblique cut, extending from the centre of the sigmoid notch to the concavity of the mandible, see Fig. 313, p. 471. (465). Raise the coronoid process with the attached temporal muscle, separating it from the structures immediately subjacent. Carefully reflect the muscle upward toward the insertion. In reflecting the muscle note the vessels and nerves entering its deep surface. Trace the middle and deep temporal vessels, and the deep temporal nerves; also the temporal branch of the temporo-malar nerve. To expose the structures in the pterygoid region part of the posterior portion of the mandible must be removed. Divide the neck of the mandible transversely; with the knife handle carefully separate the structures from the inner surface of the ramus as far down as the mandibular foramen; having determined the position of the foramen, divide the ramus by a horizontal incision just above that point and remove the portion of bone between the two incisions. Dissect out the fat and areolar tissue, exposing the vessels, nerves, and muscles.

Internal maxillary artery, 522–3. (514–15)
> First part of its course.
> Second part of its course.
> Third part of its course.
> Branches of the internal maxillary, 523–4. (515)
>> From the first part.
>> From the second part.
>> (From the third part.)

> *Branches of the First Part of the Internal Maxillary Artery, 524–5.*
> *(515–16–17)*

>> Deep auricular.
>> Tympanic branch, or Glaserian artery.
>> Middle or great meningeal.
>> Mandibular artery :—
>> (Incisive.)
>> (Mental.)
>> Lingual or gustatory branch.
>> Mylo-hyoidean branch.
>> Small meningeal.

> *Branches of the Second Part of the Internal Maxillary Artery, 526.* *(517)*

>> Masseteric branch.
>> Posterior deep temporal.
>> Internal pterygoid branches.
>> External pterygoid branches.
>> Buccal branch.
>> Anterior deep temporal branch.

But one branch of the third portion of the internal maxillary artery can be exposed at the present stage of the dissection—the posterior dental or alveolar branch.

> Posterior dental, or alveolar branch, 526. (517)
> Branches :—Antral.
>> Dental.
>> Alveolar, or gingival.
>> Buccal.

Veins, 653. (640)
    Pterygoid plexus :—
        Middle meningeal.   Posterior dental vein.   Mandibular.   Masseteric.
        Buccal.   Pterygoid veins.   Deep temporal.   Spheno-palatine vein.
        Supraorbital.   Superior palatine.   Ophthalmic.   Vesalian vein.
    Internal maxillary vein.
Pterygoideus externus, 470–1–2.   (464–5)
    Origin :—Upper head.
             Lower head.
    Insertion :—Upper head.
             Lower head.
    Structure.   Nerve-supply.   Action.   Relations.   Variations.
The temporo-mandibular articulation, 189–90–1–2.   (198–9–200–1)

        Class :—*Diarthrosis.*   Subdivision :—*Ginglymo-Arthrodia.*

    Capsular ligament :—
        Anterior portion.
        Posterior portion.
        External portion, or external lateral ligament.
        Internal portion, or short internal lateral ligament.

Remove the external portion of the capsular ligament, and expose the interior of the joint and the interarticular cartilage.

    Interarticular cartilage.
    Two synovial membranes.
    Spheno-mandibular ligament.
    Stylo-mandibular ligament.
    Arterial supply.
    Nerves.
    Movements :—
        Ginglymoid, or hinge character.
        Horizontal gliding action.
        Oblique rotary.

Divide the ligaments of the temporo-mandibular articulation and throw forward the articular process, with the attached insertions of the external pterygoid ; in doing this, special care must be exercised not to injure structures immediately internal to the joint. The small branches of the first portion of the internal maxillary artery can now be traced to the base of the skull. The mandibular division of the fifth nerve should be exposed and its branches traced.

Third or mandibular division of the fifth nerve, 801–2–3–4.   (783–4–5–6)
    Recurrent nerve.
    Nerve to the internal pterygoid.
    Anterior portion :—
        Temporal nerves.
        Masseteric nerve.
        Nerve to the external pterygoid.
        Long buccal nerve.
    Posterior portion :—
        Auriculo-temporal nerve :—
            Articular branch.
            Nerves to the meatus.
            Parotid branches.
        Mandibular nerve :—
            Mylo-hyoid.
            (Alveolar branches.)
            (Incisive branch.)
        Lingual nerve.
Chorda tympani, 808.   (790)
Pterygoideus internus, 472.   (465–6)
    Origin.   Insertion.   Structure.   Nerve-supply.   Action.   Relations.

Lymphatics from the temporal and zygomatic fossæ, 691. (676)

The structures in the submaxillary or supra-hyoid triangle will now be exposed. Divide the facial artery and vein where they cross the mandible. Draw the tongue forward and stitch it to the nose. Divide the anterior belly of the digastric at its origin from the mandible and throw it back, exposing the mylo-hyoid.

Submaxillary or Supra-hyoid region.
Submaxillary lymphatic glands, 690. (675)
    Supra-hyoid.
Lingual-glands, 692. (677)
Mylo-hyoid, 479–80. (472–3)
    Origin. Insertion. Structure. Nerve-supply. Action. Relations. Variations.

Divide the mylo-hyoid at its origin, beginning at the outer border; reflect the muscle inward and downward to the insertion, detaching it along the median raphe. With the saw, divide the mandible about a quarter of an inch external to the symphysis. Tilt the lateral portion of the mandible upward from the lower border and retain it in that position while the structures of the region are being exposed. Care must be used not to injure the mucous membrane of the floor of the mouth. The submaxillary gland is now exposed, with the surrounding muscles and the structures crossing them. The submandibular ganglion lies between the submaxillary gland and the hyo-glossus muscle; it can be readily exposed by carefully turning down the gland from its upper border and following the chorda tympani and the branches of the lingual nerve passing to it. The stylo-glossus, hyo-glossus, and genio-hyoid muscles will be exposed, also the anterior portion of the genio-hyo-glossus.

Submaxillary gland, 984. (963)
    Duct of the submaxillary gland.
    Arteries.
    Nerves.
Sublingual gland, 984–5. (963)
    Duct of Rivini.
    Arteries.
    Nerves.
Lingual nerve, 804–5. (786)
    Branches.
Submandibular ganglion, 805. (787)
    Branches.
Hypoglossal nerve, 821–2. (802)
    Nerve to the genio-hyoid.
    True hypoglossal branches :—
        Nerve to the stylo-hyoid.
        Nerves to the hyo-glossus.
        Nerves to the genio-hyo-glossus.
Stylo-glossus, 482–3. (474–5)
    Origin. Insertion. Structure. Nerve-supply. Action. Relations.
Hyo-glossus, 481–2. (474)
    Origin. Insertion. Structure. Nerve-supply. Action. Relations.
Genio-hyoid, 480. (473)
    Origin. Insertion. Structure. Nerve-supply. Action. Relations. Variations.

Divide the hyo-glossus at its origin; carefully separate it from the subjacent structures; draw it from under the structures crossing its surface; throw it upward and expose the structures beneath it.

Lingual artery, 513–14–15. (505–6)
    Second part of its course.
    Third part of its course.
    Branches :—Dorsalis linguæ.
            Sublingual artery :—
                Artery of the frænum.
            Ranine artery.
Lingual vein, 665–6. (652)
Genio-hyo-glossus, 481. (473–4)
    Origin. Insertion. Structure. Nerve-supply. Action. Relations.

Divide the posterior belly of the digastric at its origin and throw it downward; in raising the muscle note the branch of the facial nerve which supplies it, also the branch to the stylo-hyoid. Divide the external carotid just below its termination; cut its posterior branches at their origin, and throw the vessel forward. The deep structures can now be exposed.

Nerve to the posterior belly of the digastric, 809. (790)
Nerve to the stylo-hyoid, 809. (790)
Stylo-pharyngeus, 987. (965)

Deep Dissection of the Neck.

Note the position of the glosso-pharyngeal nerve and its relation to the stylo-pharyngeus muscle. Divide the styloid process at its base and throw it forward and downward with its attached muscles. Expose the upper part of the internal jugular vein; then divide the vein about two inches below its commencement and throw it upward. Carefully expose the internal carotid artery and the nerves in relation with it. Small nerve branches marked (—)*, can only be demonstrated by a special dissection on a fresh subject and need not be looked for.

Internal jugular vein; bulb, 665. (651–2)
    Tributaries.
Internal carotid artery, 528–9. (519–20)
    The cervical portion :—
        Relations :—Behind.
                Outer side.
                Inner side.
Ascending pharyngeal artery, 510–11–12. (502–3)
    Branches:—Prevertebral.
            Pharyngeal.
            Palatine.
            Meningeal.
Glosso-pharyngeal nerve, 814–15. (795–6)
    " From its superficial origin."
    (Jugular ganglion.)*
    (Petrous ganglion.)*
    Branches :—(Meningeal branches.)*
            (Tympanic branch.)*
                Communicating twigs :—
                    (a), (b), (c), (d), (e), (f).
            (Communicating branches.)*
                See—lingual branch, 809. (790)
            Muscular branch.
            Pharyngeal branches.
            Tonsillar branches.
            Lingual branches.

The cervical portion of the pneumogastric may now be traced and its branches followed.

Pneumogastric or vagus nerve, 815–16–17–18. (798–9–800)
    " From its superficial origin."
    (Ganglion of the root.)*
    Ganglion of the trunk :—
        Branches:—Communicating branches :—
                (a), (b), (c), (d), (e).
        Branches of distribution :—
            (Meningeal or recurrent branch.)*
            (Auricular branch, or nerve of Arnold.)*
            Pharyngeal branches.
            Superior laryngeal nerve :—
                Internal branch.
                External laryngeal branch.
            Cardiac branches.
Spinal accessory nerve, 820. (801)
    " At the base of the skull."
Hypoglossal nerve, 821. (802)
    " The filaments unite to form two fasciculi, etc."
    (Meningeal branch.)*

Sympathetic nerves, 863-4.   (843-4)
Gangliated cords of the sympathetic, 864-5-6-7-8.   (844-5-6-7)
    Cervical portion of the gangliated cord :—
        Superior cervical ganglion, rami communicantes.
            Branches:—Ascending branch.
                  (Carotid plexus) :—
                      Tympanic branch.   Great deep petrosal nerve.
                      Branches to the Gasserian ganglion.
                      Branches to the sixth nerve.
                  (Cavernous plexus) :—
                      Communicating branches to the third, fourth,
                      and ophthalmic division of the fifth cranial nerves.
                      Sympathetic root of the lenticular ganglion.
            Nervi molles.
            Communicating branches to the cranial nerves.
            Pharyngeal branches.
            Superior cervical cardiac nerve.
        Middle cervical ganglion ; rami communicantes ; ansa Vieussenii.
            Branches:—Branches to the thyroid body.
                Middle cervical nerve.
      Inferior cervical ganglion, rami communicantes.
            Branches:—Branches to the vertebral artery.
                Inferior cardiac nerve.
Relations—of the pharynx, 987.   (965-6)

Prevertebral Region.   Articulations.

Remove the sterno-mastoid, sterno-hyoid, sterno-thyroid, and the thyroid body.   The structure of the pharynx can be demonstrated to advantage only when separated from the spine and posterior portion of the skull; it is therefore necessary to divide the skull and remove the anterior portion with the pharynx attached.   Divide the trachea, œsophagus, and recurrent laryngeal nerves about an inch below the larynx ; cut the common carotid artery about an inch below its termination ; divide the vagus and sympathetic nerves at the same level.   Draw the trachea and œsophagus forward, carefully separating the pharyngeal walls from their prevertebral attachment.   At the base of the skull, divide the periosteum and expose the basilar process between the pharynx and the prevertebral muscles; with the chisel, divide the basilar process at this point, driving the chisel upward through the bone.   With a narrow saw, carry a cut inward on each side, along the posterior border of the petrous portion of the temporal bone, passing just behind the jugular foramen, then curving forward on its inner side to the suture between the basilar process and the petrous portion of the temporal bone, joining the first cut, made with the chisel.   Great care must be exercised not to injure the pharyngeal wall.   The pharynx should be wrapped in a cloth, moistened with preservative fluid, while the remaining structures of the prevertebral region are exposed.

Longus colli, 486-7.   (478)
      Vertical portion :—
          Origin.
          Insertion.
      Lower oblique portion :—
          Origin.
          Insertion.
      Upper oblique portion :—
          Origin.
          Insertion.
      Structure.   Nerve-supply.   Action.   Relations.
Rectus capitis anticus major, 485.   (477)
      Origin.   Insertion.   Structure.   Nerve-supply.   Action.   Relations.
Rectus capitis anticus minor, 485-6.   (478)
      Origin.   Insertion.   Structure.   Nerve-supply.   Action.   Relations.
Intertransversales, 449-50.   (445)
      Structure.   Nerve-supply.   Action.   Relations.

Remove the muscles covering it and expose the vertebral portion of the vertebral artery.

Vertebral artery :—
      Second or vertebral portion, 540.   (532)

Branches of the second or vertebral portion, 542.
Lateral spinal branches.
Muscular branches.
The Ligaments and Joints between the Skull and Spinal Column, and between the Atlas and Axis, 193–200. (201–208)
The Articulation of the Atlas with the Occiput.
Class:—*Diarthrosis.* Subdivision:—*Ginglymo-Arthrodia.*
Anterior occipito-atlantal ligament.
Posterior occipito-atlantal ligament.
Capsular ligaments.
Anterior oblique or lateral occipito-atlantal ligament.
Synovial membrane.
Arterial supply.
Nerve-supply.
Movements:—Directly lateral.
Obliquely lateral.
The Articulations between the Atlas and Axis.
The Lateral Atlanto-axoidean Joints. { Class:—*Diarthrosis.* Subdivision:—*Arthrodia.*
The Central Atlanto-axoidean Joint. { Class:—*Diarthrosis.* Subdivision:—*Trochoides.*
Anterior atlanto-axoidean ligament.
Posterior atlanto-axoidean ligament.
The Lateral Atlanto-axoidean Joints:—
Capsular ligaments.
Synovial sac.
The Central Atlanto-axoidean Joint:—
Transverse ligament. Crucial ligament.
Atlanto-odontoid capsular ligament.
Synovial membranes.
Arterial supply.
Nerve supply.
Movements.
The Ligaments uniting the Occiput and Axis.
Occipito-cervical or cervico-basilar ligament.
Lateral occipito-odontoid or check ligaments.
Central odontoid or suspensory ligament.

Dissection of the Pharynx.

The dissection of the pharynx and the structures of the anterior portion of the skull will now be continued; first examine the cavity of the mouth, and the palate, through the buccal orifice.

The mouth, 979–80. (958–9)
Fauces.
Vestibule.
Buccal orifice:—
Upper and lower lips:—
Lips.
Cheeks.
Gums.
The palate, 980. (959)
Hard palate.
Soft palate:—
Uvula.
Pillars of the fauces:—
Anterior pillar.
Posterior pillar.
Tonsillar recess.
Isthmus of the fauces.
Anterior surface of the soft palate.

Moderately distend the pharynx with tow; place it with its posterior surface uppermost, and demonstrate the muscles forming its wall.

The pharynx :—
  Pharyngeal walls, 985. (964)
      Veins of the pharynx, 665. (651)
      Muscles, 985-6-7. (964-5)
          Muscular coat :—Constrictor muscles.
                          Stylo- and palato-pharyngei.
              Inferior constrictor.
              Middle constrictor.

              Divide the internal pterygoid transversely near the middle, and draw aside its
              upper and lower portions to fully expose the origin of the superior constrictor and
              the middle portion of the buccinator.

              Superior constrictor. Sinus of Morgagni.

                  Note the continuity of the superior constrictor with the middle portion of
                  the buccinator, through the pterygo-mandibular ligament.

                  Buccinator, 462-4. (457-8)
                      Origin, (2). Relations.
              Stylo-pharyngeus, 987. (965)
      (Palato-pharyngeus.)
      Pharyngeal aponeurosis.

Open the pharynx by a median incision through its posterior wall, extending from the base of the
skull to the commencement of the œsophagus ; at the base of the skull make a lateral incision each
way from the median incision.

The pharynx, 985. (964)
      Nasal portion.
      Buccal portion.
      Mucous membrane.
      Interior of the pharynx, 987. (966)
          Pharyngeal recess.
          Pharyngeal bursa.
          Pharyngeal tonsil.
          Eustachian tube, 913. (892)
              Cartilaginous portion.
              Mucous membrane.
          Lymphatics of the pharynx, 692. (676)
          Internal maxillary, or deep facial glands, 692. (677)
          Post-pharyngeal gland.

The soft palate and associated parts should now be examined from behind.

The soft palate, 980-1-2. (959-60-61)
      Uvula.
      Pillars of the fauces:—
          Anterior pillar.
          Posterior pillar.
      Tonsillar recess.
      Isthmus of the fauces.
      Mucous membrane of the soft palate.

Strip off the mucous membrane and expose the muscles of the soft palate.

      Muscles:—
          Levator palati.
          Azygos uvulæ.
          Palato pharyngeus.
          Palato glossus.
          Tensor palati.
      Arterial supply of the soft palate.
      Nerves to the soft palate.

Tonsils :—
    Arteries of the tonsil.
    Veins of the tonsil.
    Nerves of the tonsil.

Beginning at the angle of the mouth, make an incision backward, dividing the buccinator, mucous membrane, pterygo-maxillary ligament, and the superior constrictor. Wrap the anterior portion of the skull in a cloth moistened with preservative fluid and lay it aside until the dissections of the tongue and the larynx are completed. The tongue will be examined first.

The tongue, 921-2-3-4-5. (900-1-2-4)
    Special sense of taste. Function of speech. Mastication and deglutition.
    Dorsum. Tip.
    Mucous membrane :—
        Folds :—Glosso-epiglottidean.
            Frænum epiglottidis.
                Glosso-epiglottidean pouch, or vallecula.
        Anterior pillars of the fauces.
        Frænum linguæ.
    Median raphe. Foramen cæcum.
    Papillæ :—
        Circumvallate, or caliciform papillæ.
        Fungiform papillæ.
        Filiform, or conical papillæ.
    Lingual glands.
    Gland of Nuhn.
    Lymphoid tissue.
    Muscles :—
        Extrinsic muscles.
        The intrinsic muscles :—
            Lingualis superior.
            Inferior lingualis.
            Transverse fibres.
            Vertical fibres.
    Fibrous septum. Hypoglossal membrane.
    Arteries.
    Nerves :—
        Mandibular division of the fifth.
        Glosso-pharyngeal.
        Superior laryngeal.
        Hypoglossal.
        Chorda tympani.
    The lymphatics of the mouth and tongue, 691. (676)

The larynx will now be considered. With prepared specimens at hand, read the description of the cartilages of the larynx, and study their arrangement and relations. The cartilages of the specimen undergoing dissection will be gradually exposed as the dissection proceeds.

Larynx, 938. (917)
    Cartilages :—
        Single cartilages.
        Paired cartilages.
        Hyaline.
        Yellow elastic.
        Thyroid cartilage, 939-40. (918-19)
            Outer surface.
            Inner surface.
            Anterior, or isthmic border, pomum Adami.
            Superior border.
            Inferior border.
            Superior cornua.

Thyroid cartilage :—
    Inferior cornua.
Cricoid cartilage, 940–1. (919–20)
    Anterior and posterior portion.   Superior and inferior border.
    Posterior quadrate portion.
    Anterior portion.
    Inner surface.
    Superior border.
    Inferior border.
Arytenoid cartilages, 941. (920)
    Three surfaces. Three borders.
    Base. Apex. Angles.
    Surfaces :—
        Posterior. Anterior. Internal.
    Base.
    Angles :—
        Anterior. External, or muscular. Internal.
    Borders :—
        Internal. External.
    Apex.
Cornicula laryngis, or cartilages of Santorini, 942. (921)
Cuneiform cartilages, or cartilages of Wrisberg, 943. (921)
Calcification. Cartilage triticea.
Joints of the larynx, 944–5. (923)
    Crico-thyroid joints.
    Crico-arytenoid joints.
        Posterior crico-arytenoid ligament.
        Crico-arytenoid ligament.
    " Cricoid articular surface."
    " The arytenoid."
        Rotates.
        Glides.

A general examination should now be made of the specimen to be dissected, and the arrangement of its parts studied, so far as is possible, without dissection. Fig. 507, P. 922 (901) shows the parts displayed by an examination of the upper extremity of the larynx, also Figs. 533–4.

The epiglottis, 941–2. (920–1)
    Glosso-epiglottidean folds.
    Cushion of the epiglottis.
    Aryteno-epiglottidean folds.
    Pharyngo-epiglottidean fold.
The interior of the larynx, 947–8–9–50. (926–7–8–9)
    Superior aperture, or opening of the glottis. Sinus pyriformis.
    Mucous membrane.
    Suprarimal portion :—
        Ventricle.
        Laryngeal pouch.
        (Compressor sacculi laryngis.)
        Superior or false vocal cords. Fossa innominata.
    Rima glottidis.
    Infrarimal portion.

Dissection of the Larynx.

Sever the tongue from the hyoid bone. Remove the fibres of origin of the inferior constrictor. Expose the thyro-hyoid membrane and ligaments and clean the superficial muscles of the larynx, carefully preserving the laryngeal nerves and vessels.

Thyro-hyoid membrane. Thyro-hyoid ligaments, 943. (922)

Demonstrate the crico-thyroid muscles and expose the portion of the crico-thyroid membrane between the muscles.

Crico-thyroid, 945 ; Fig. 537. (924)

Crico-thyroid membrane, 943. (922)
Crico-arytenoideus posticus, 945; Fig. 535. (924)

Divide the thyroid cartilage by a vertical incision just to the right of the anterior median line; carefully raise the right portion of the cartilage; divide the ligaments and remove it, exposing the structures underneath.

Crico-arytenoideus lateralis, 945-6; Fig. 528. (924)
Thyro-arytenoideus muscle, 946-7; Fig. 528. (524-5)
Arytenoideus, 947; Fig. 528. (925-6)
    Aryteno-epiglottideus muscle.
Thyro-epiglottideus, 947; Fig. 528. (926)

Carefully remove the crico-arytenoideus lateralis and the thyro-arytenoideus, and expose the outer surface of the lateral portion of the crico-thyroid membrane and thyro-arytenoid ligaments.

Crico-thyroid membrane, 943-4. (922-3)
    True vocal cords or inferior thyro-arytenoid ligaments.
Superior thyro-arytenoid ligaments, 944, (923)

Divide the crico-thyroid membrane and vocal cords on the right side by a vertical incision in the median line of the lateral wall; the interior of the larynx will now be exposed, and the vocal cords, etc., of the left side should now be studied from within. The nerves and vessels can then be followed on the left side.

Nerves, 950; Fig. 535. (928-9)
    The superior laryngeal.
    Inferior laryngeal :—
        Anterior branch.
        Posterior branch.
Arteries. Veins. Lymphatics, 950. (929)
Lymphatics of the larynx, 692. (676)
    Laryngeal glands.
Hyo-epiglottidean ligament. Periglottis, 942. (921)
Epiglottis, 941-2. (920)

Internal Orbital Region.

With the saw cut through the frontal bone and the plate forming the roof of the orbit. Make two incisions; the outer one extending from the outer end of the sphenoidal fissure to the outer angle of the orbit, and the inner one from a point just external to the optic foramen to the inner side of the orbit, but external to the pulley of the superior oblique muscle. Beginning at the posterior border, raise and tilt forward this triangular plate, exposing the orbital periosteum underneath.

Fascia of the orbit, (1), (2), (3), 893. (873)
    Orbital periosteum, or periorbita, 894. (874)

Divide the periosteum in the middle line of the orbit, and transversely at the anterior border of the orbit. Reflect the fascia each way from the median incision and expose the structures within.

Cavity of the orbit, general arrangement of its contents, 890-1. (870)
The frontal nerve, 796. (778)

Trace the supraorbital and supratrochlear branches to the point of exit from the orbit.

The supraorbital artery, 531. (522)
    Branches:—Periosteal. Muscular. Diploic. Trochlear. Palpebral.
Lachrymal nerve, 796. (778)
Lachrymal artery, 530-1. (521-2)
    Recurrent lachrymal. Muscular. Malar branches. Palpebral branches.
The lachrymal gland, 904-5. (883-4)

Carefully raise the gland and tease out the thread-like ducts passing from the gland to the conjunctival sac.

Nerves of the orbit, 900-1. (879-80)
    Motor. Sensory. Sympathetic.
        A. Motor nerves.
        B. Sensory nerves.
        C. Sympathetic nerves of the orbit.
Fourth or trochlear nerve, 793-4, 900. (775-6, 880)

4

Muscles of the orbit, 891–2. (870–1)
    (Superior and inferior recti. Superior and inferior obliques.)
    (Internal and external recti. Levator palpebræ superioris.)
Fascia of the orbital muscles, 2, 894–5–6. (875–6)

The sheaths of the orbital muscles and the check ligaments will be developed as the dissection proceeds.

Levator palpebræ superioris, 892. (872)

Divide the frontal nerve and turn it forward, also divide the levator palpebræ superioris midway between the origin and insertion, reflecting the anterior portion forward and the posterior portion to the origin. The eyeball should be distended; to do this, make a small oblique puncture at the junction of the sclerotic and cornea, insert the tip of a small blowpipe, and inflate the eyeball. The superior rectus should then be exposed.

The superior rectus (four recti muscles), 892. (871–2)
The superior oblique, trochlea, 892–3. (872–3)

Divide the superior rectus in the same manner as the levator palpebræ superioris and reflect the divisions. Carefully remove the fat and fascia and expose the structures posterior to the eyeball, noting the plan of formation of the sheaths of the muscles as they are exposed. The optic nerve will be exposed, and, crossing the optic nerve, the nasal nerve and the ophthalmic artery and vein. On the outer side of the optic nerve, close to the ophthalmic artery, is the lenticular ganglion; the roots of the ganglion and the branches from it should be carefully traced.

The nasal nerve, 796–7, 901. (778–9, 880)
    Branches:—Long root to the lenticular ganglion.
             Long ciliary nerves.
             Infratrochlear nerve.
Lenticular, ciliary, or ophthalmic ganglion, 797, 901. (779, 881)
    Motor or short root. Sensory or long root.
    Sympathetic root. Short ciliary nerves.
Optic nerve, 791, 897–8. (773, 877–8)
    Dural sheath, pial sheath.
Ophthalmic artery, 530–1–2–3, 899. (521–2–3–4, 879)
    Branches of the ophthalmic artery:—
        Arteria centralis retinæ.
        Muscular branches.
        Ciliary arteries:—
            Short posterior.
            Long posterior.
            Anterior.
        Posterior ethmoidal:—
            Ethmoidal branches.
            Meningeal branches.
            Nasal branches.
        Anterior ethmoidal:—
            Ethmoidal. Meningeal. Nasal.
            Frontal. Cutaneous or terminal.
        Palpebral. Frontal. Nasal. Lachrymal. Transverse nasal.
Veins of the orbit, 663–4–5, 900. (650–1, 879)
    Ophthalmic vein, or common ophthalmic vein.
        Superior ophthalmic vein.
            Tributaries:—
                Superior muscular branches.
                Ciliary veins:—
                    Anterior and posterior set.
                Anterior and posterior ethmoidal veins.
                Lachrymal vein.
                Central vein of the retina.
        Inferior ophthalmic vein.
            Tributaries:—
                Inferior muscular.
                Lower posterior ciliary veins.

The lymphatic system of the orbit, 901. (881)
Third nerve, 791-2-3, 900. (774-5, 880)
    Superior division.   Inferior division.
Sixth nerve, 805-6, 900. (787-8, 880)
The (four) recti muscles, 892. (871-2)
    "Internal rectus.   Inferior rectus.   External rectus."
Tenon's capsule.   Tenon's space, 897. (876)
Inferior oblique, 893. (872)
Relation of Tenon's capsule to the oblique muscles, 897. (876)
Action of the ocular muscles, 893. (872-3)
    Principal axes.
        Abduction and adduction.
        Rotation of the cornea.
    Action of :—
        External rectus.   Internal rectus.
        Superior rectus.   Inferior rectus.
        Superior oblique.   Inferior oblique.

Divide the muscles and the optic nerve, and remove the eyeball from the optic. Note the form of the eyeball, and expose the insertion of the ocular muscles. The orbital nerve should be traced, and the periosteum of the lower portion of the orbit examined.

Orbital or temporo-malar nerve, 798-9. (780-1)
    Temporal branch.   Malar branch (orbital portion).
Orbital periosteum or periorbita, orbital muscle, 894. (874)
Examination of the eyeball, 878-9-80. (858-9-60)
    Anterior and posterior pole.
    Sagittal axis.
    Equator.

Otic Ganglion. Maxillary Division of the Fifth Nerve, etc.

Expose the otic ganglion ; to find the ganglion, follow the nerve to the internal pterygoid, 802. (783-4) —toward its origin, near its commencement, it gives off a branch which can be traced to the ganglion.

Otic ganglion, 805. (787)
    Motor root.
    Sympathetic root.
    Branches.

The trunk of the maxillary division of the fifth nerve should now be exposed. With the saw, divide the malar process of the maxilla by a cut just external to the infraorbital groove ; make a second cut from the root of the zygoma inward and forward to the middle of the sphenoidal fissure and remove the intermediate portions of bone. With the chisel, chip away the bone and expose the nerve and its branches in place.

Maxillary division of the fifth nerve, 797-8. (780)
    Posterior superior dental nerves, 799. (781)
    Middle and anterior superior dental nerves, 799. (781)
        Middle dental nerve.
        Anterior dental nerve :—
            Nasal branch.
            Ganglion of Valentin.
            Ganglion of Bochdalek.
Internal maxillary artery :—
    Infraorbital branch, 526-7. (517-18)
        Orbital.
        Anterior dental branch.
        Nasal branches.

With the saw divide the anterior portion of the skull into two lateral parts by a sagittal cut passing just to one side of the nasal septum. The septum must not be injured.

The Nasal Fossæ. Meckel's Ganglion, etc.

The mucous membrane, etc., 930-1-3. (909-10-12)
    Vestibule.   Roof.   Floor.

The Nasal Fossæ—Mucous Membrane, etc.:—
>   Superior meatus.
>   Middle meatus.
>   Inferior meatus.

Expose and trace the nerves and vessels; first, those of the septum; then remove the septum and trace the nerves and vessels of the lateral walls. The cartilage of the septum should be examined and its connections noted before it is removed.

Septal cartilage, 928.   (907)
>   Anterior border.
>   Posterior border.
>   Inferior.

The nerves, 933-4.   (912)
>   Olfactory, or special nerves of smell:—
>>   Posterior. Anterior.
>   Nasal branch of the ophthalmic.
>   Vidian, with the upper anterior branches of Meckel's ganglion.
>   Naso-palatine.
>   Anterior dental branch of the maxillary division of the fifth nerve.
>   Anterior, or large palatine nerve.

Arteries, Veins, 934.   (912-13)
>   Spheno-palatine:—
>>   Internal, or naso-palatine branch:—
>>   Supplies—the septum.
>>   External branch:—
>>   Supplies—antrum, frontal sinus, ethmoidal cells, turbinated bones
>>   meatuses.
>   Anterior and posterior ethmoidal arteries:—
>>   Supply—septum, roof, outer wall, posterior ethmoidal cells.
>   Descending palatine:—
>>   Branches to—inferior meatus and lower turbinal bone.

Veins:—
>   Spheno-palatine, ethmoidal, alveolar.
>   Communications.

Lymphatics.   Lymphatics from the interior of the nose, 691.   (676)

With the chisel carefully chip away the inner wall of the posterior palatine canal, and expose Meckel's ganglion and the branches to and from it.

Spheno-palatine, nasal, or Meckel's ganglion, 800-1.   (782-3)
>   Descending branches:—
>>   Great, or anterior palatine:—
>>>   (Inferior nasal nerves.)
>>   Posterior, or small palatine.
>>   External palatine nerve.
>   Posterior branch.
>   Pharyngeal branch.
>   (Internal branches):—
>>   Superior nasal.
>>   Septal branches:—Naso-palatine nerve, or nerve of Cotunnius.
>   Ascending, or orbital branches.

Branches of the third part of the internal maxillary artery, 526-7.   (517-18)
>   (Posterior dental, or alveolar):—
>>   Antral. Dental. Alveolar, or gingival. Buccal.
>   (Infraorbital):—
>>   Orbital. Anterior dental branch. Nasal branches.
>   Posterior, or descending palatine branch:—
>>   Anterior branch.
>>   Posterior branch.
>   Vidian artery:—
>>   Pharyngeal. Eustachian. Tympanic.

Pterygo-palatine artery, or pterygo-pharyngeal :—
 Pharyngeal. Eustachian. Sphenoidal.
Spheno-palatine, or naso-palatine artery :—
 Pharyngeal branch. Sphenoidal branch. Nasal branches.
 Ascending septal branches.

With the chisel chip away the wall and expose the lachrymal sac and the nasal duct.

Lachrymal sac. Nasal duct. Lachrymal canal, 905–6. (884–5)

With the chisel, chip away the anterior inferior wall of the carotid canal, and expose the intraosseous portion of the internal carotid.

Intraosseous portion, 529. (520)
 Branches of the intraosseous portion :—
  Tympanic.
  Vidian branch.

Middle Ear. Internal Ear.

Remove the temporal bone, with the pinna. Cut away the squamous portion above the zygoma. With a chisel, chip away the roof and anterior wall of the external auditory meatus, exposing the canal and the outer surface of the membrana tympani.

External Auditory Meatus, 909–10. (888–9)
 Cartilaginous portion.
 Osseous portion.
 Lining membrane.
 Arteries.
 Nerves.

Carefully chip away the tegmen tympani with a small, sharp chisel, and expose the tympanic cavity.

The Middle Ear, 910–916. (889–894)
 Membrana tympani. Rivinian segment. Umbo.
  Structure :—Fibrous layer, external and internal lamellæ.
   Cuticular covering.
   Mucous lining.
 Tympanic cavity :—
  Roof. Floor. Outer wall.
  Chorda tympani nerve, Iter chordæ posterius, Iter chordæ anterius.
 Inner wall :—Fenestra ovalis, fenestra rotunda, Fallopian canal.
  Fenestra ovalis.
  Fenestra rotunda :—
   Membrane of:—Middle fibrous layer.
    Mucous lining.
    Serous layer.
  Posterior wall.
  Anterior wall.
 Eustachian tube :—
  Osseous portion.
  Cartilaginous portion.
  Mucous membrane.
 Ossicles of the ear, 73-4-5. (85-6-7)
  Malleus :—
   Head. Neck. Handle, or manubrium. Slender process. Short process.
  Incus :—
   Body. Short process. Long process. Orbicular tubercle.
  Stapes :—
   Head. Base. Crura. Neck.
 Articulation of the ossicles, 913, etc. (892, etc.)
  The malleus with the incus. Capsule. Meniscus.
  Incus with the stapes.
   Elastic capsular ligament.
  Stapes with the margin of the fenestra ovalis.

Ligaments of the ossicles :—
    Superior ligament of the malleus.
    Anterior ligament of the malleus.
    External ligament of the malleus.
    Ligament of the incus.
Muscles of the tympanum :—
    Tensor tympani ; arises, inserted.
    Stapedius ; arises, inserted.
      Nerves.
Mucous membrane of the tympanum :—
    (*a*).
    (*b*) Obturator ligament of the stapes.
    (*c*).
Vessels and nerves of the tympanum.
    Arteries :—
      Tympanic branch.   Stylo-mastoid branch.   Petrosal branch.
    Veins.
    Nerves :—
      Tympanic plexus :—
        Tympanic branch of glosso-pharyngeal.
        Communicating branch from carotid plexus of sympathetic.
        Great superficial petrosal.
        Small superficial petrosal.

With a fine saw divide the petrous portion of a temporal bone by a coronal incision passing through the external and internal auditory canals. Also divide the petrous portion of a temporal bone by a vertical incision, from the superior border downward through the middle of the mastoid process. Some idea of the internal ear can be obtained by careful study of these sections. This should be supplemented by the study of specially prepared specimens, in which the osseous labyrinth has been exposed, and by models and diagrams.

Internal Ear, or Labyrinth, 916–17–18–19–20. (895–6–7–8–9)
    Osseous labyrinth :—
      Vestibule. Semicircular canals. Cochlea.
      Vestibule :—
        Outer wall, fenestra ovalis.
        Inner wall, fovea hemispherica, aqueductus vestibuli.
        Roof, fovea hemielliptica.
        Openings :—Behind.
          In front, apertura scalæ vestibuli, by which vestibular cavity communicates with the scala vestibuli of the cochlea.
      Three semicircular Canals, ampulla.
        Superior semicircular canal.
        Posterior semicircular canal.
        External semicircular canal.
      Cochlea :—
        Base ; tractus spiralis foraminulentus.
          Modiolus, or columella :—
            Spiral canal, lamina spiralis, hamulus, heliocotrema.
              Lamina spiralis.
              Scala vestibuli and scala tympani.
                Scala vestibuli.
                Scala tympani.
          Central axis, or modiolus ; cupula, infundibulum.
            Central canal of the modiolus.
    Membranous labyrinth :—
      Perilymph. Canal of the cochlea.
      Within the vestibule :—
        Utricle and saccule.
        Utricle.

Membranous Labyrinth—Within the Vestibule :—
   Saccule. Canalis reuniens.
  Within each semicircular canal.
  Structure :—Tunica propria.
    Fibrous investment.
    Epithelial lining.
  Within the cochlea :—
    Membrana basilaris.
    Membrane of Reissner.
     Ductus cochlearis, or membranous canal of the cochlea. '
     Canalis reuniens.
Auditory nerve :—
  Superior division : —
    To utricle, superior and external semicircular canals.
  Inferior division :—
    To cochlea, saccule, posterior semicircular canal.
Vessels : —
  Artery :—Internal auditory branch :—
    To vestibule and cochlea, with their membranous contents.
  Veins.

With a small chisel chip away the bone and expose the canal for the facial nerve.

Facial nerve, 807–8. (788–9)
  "At its superficial origin." Geniculate ganglion.
  Branches, (1) :—
    Great superficial petrosal nerve.
    Lesser superficial nerve.
    External superficial petrosal.
    Nerve to the stapedius.
    Chordi tympani.
    Communicating twig to the pneumogastric.

Dissection of the Eyeball.

For the dissection of the eyeball, procure several eyes of the pig, sheep, or ox. The eye of the pig most closely resembles the human eye in form and structure, but on account of its size the eye of the ox is more readily dissected. The conjunctiva, ocular muscles, and fat should be carefully removed, exposing the eyeball; in removing these, note the vena vorticosæ issuing from the sclerotic, a little posterior to the equator of the eyeball ; and near the optic nerve, the ciliary vessels and nerves entering the eyeball through the sclerotic.

To obtain a general idea of the arrangement of the parts composing the eyeball and for reference during the dissection, make a series of sections of the eyeball ; to do this put a couple of eyes in small tin boxes and place in a mixture of salt and crushed ice for three or four hours till frozen solid. With a fine saw divide one of the eyes into an anterior and a posterior section, cutting it at the equator of the eyeball; divide the other eye into two lateral halves by an antero-posterior section, cut one of these halves into a number of thin sections. Pin the sections in a tray and cover with water ; they can then be examined to advantage with a glass.

General structure of the eyeball.

  On examination of the sections it will be seen that the eyeball consists of tunics and of refracting media or humors :—
   The tunics forming the coats of the eyeball are :—
    1. The sclerotic and cornea.
    2. The choroid, ciliary processes and iris.
    3. The retina.
   The refracting media or humors are :—
    1. The vitreous humor.
    2. The crystalline lens and capsule.
    3. The aqueous humor.

Divide an eye at the equator of the eyeball and note the parts seen in each hemisphere.

1. Posterior hemisphere, 880. (860)
  Retina. Retinal pigment layer. Membrana suprachoroidea.

2. Anterior hemisphere, 880–1. (860–1)
    Crystalline lens. Ora serrata. Iris. Lens capsule.
    Suspensory ligament. Ciliary muscle.

The coats of the eye should now be systematically demonstrated on a fresh eye.

The outer, fibrous coat of the eye, 881. (861)
    Lamina cribrosa, limbus. Lamina cribrosa, 898–9. (878–9)

With a sharp knife carefully cut through the sclerotic at the equator of the eyeball, to the choroid; with a pair of sharp scissors continue the incision around the eyeball. Make three or four incisions backward nearly to the optic nerve and carefully reflect the sclerotic from the choroid. Divide the optic nerve close to the inner side of the sclerotic. In the same manner carry the dissection forward to the margin of the cornea, carefully detaching the ciliary muscle from the inner surface of the sclerotic.

The cornea, 881–2. (861–2)
Middle or vascular coat of the eye, uveal tract, 882–3–4. (862–3–4)
    The choroid:—
        Membrana suprachoroidea.
        Membrane of Bruch.
        Choroido-capillaris, Fig. 478.
    Ciliary body:—
        Corona ciliaris.
        Orbiculus ciliaris.

With a small brush wash away the pigment on one of the anterior hemispheres, exposing the ciliary processes.

    Ciliary muscle.

The arrangement of the fibres of the ciliary muscle can be studied to advantage only with properly prepared and mounted sections and a microscope.

    The iris, sphincter iridis, 883–4. (862–3–4)

Take an eyeball from which the sclerotic has been removed and placing it in water, with a fine brush carefully wash away the pigment of the choroid and expose the vessels and nerves on the surface.

Ciliary nerves of the eyeball, 886–7; Fig. 479. (867)
The ciliary system of blood-vessels, 888–9; Figs. 479 and 478. (868–9)
    (1) Short posterior ciliary arteries.
    (2) Long posterior ciliary arteries.
    (3) Anterior ciliary arteries.
    Veins:—Vena vorticosæ.
        Anterior ciliary veins.

With the eyeball under water carefully remove the choroid, ciliary processes, and iris and expose the outer surface of the retina.

The innermost or nervous coat, 884–5. (864)
    Pars ciliaris retinæ.
    The retina proper.
        Arteria centralis retinæ, 887–8. (867–8)
        Vena centralis retinæ, Fig. 478.
Aqueous humor, 886. (867)
    Anterior. Posterior aqueous chamber.
Vitreous humor, 886. (866–7)
Suspensory ligament of the lens, canal of Petit, 886. (866)
The lens, 885–6. (864–5)
    Equator. Pole. Stellate figures.
    Nucleus. Cortex.
Lymphatic system of the eyeball, 889–90. (869–70)
    Anteriorly:—
        In the cornea.
        In the iris.
    Posteriorly.

Dissection of the Brain.

Before beginning the dissection of the brain, the student should read the chapter on the mode of development—" Neurology," 708–9–10. (692–3–4)

THE BRAIN.

The Arachnoid and Pia Mater will now be considered.

The arachnoid, 715–16. (699–700)
    Pacchionian glands.

      Structure of a Pacchionian body, 718. (702)

Subarachnoid tissue, 716–17. (700–1)
    Subarachnoid spaces :—
      Cisterna magna.
      Cisterna pontis.
Pia Mater, 717. (701)
Lymphatics of the Brain and Spinal Cord, 717–18. (701–2)

    Subarachnoid space.

    Cerebro-spinal fluid.

Carefully remove the membranes from the surface of the brain. Beginning at the base, the arteries should be carefully exposed and traced as far as possible without injury to the brain substance. The main trunks can be followed, the terminal distribution of the branches will be somewhat developed during the process of the dissection. The description of the branches should be read at this time and can be reviewed later as the vessels are exposed.

Encephalon or brain, 718. (702)
    Cerebrum.   Mesencephalon.
    Cerebellum.   Peduncles or crura.

    Gray matter.   White matter.

Internal carotid.
    Intercranial portion, 529. (520)
      Anterior cerebral.   Anterior communicating artery, 534. (524–5)
        Branches of the anterior cerebral :—
          Communicating :—
            Anterior communicating.
            Antero median branches.
          Ganglionic, or central :—
            Antero-median group.
          Commissural.
          The hemispherical, or cortical branches:—
            Orbital.
            Margino-frontal.
            Calloso-marginal.
            Quadrate.
      Middle cerebral artery, 534–5. (525)
        Branches of the middle cerebral:—
          Ganglionic, or central :—
            Caudate.
            Antero-lateral.
            Lenticulo-striate.
          Hemispherical, or cortical branches :—
            Inferior, or orbito-frontal.
            Ascending frontal.
            Parietal.
            Parieto-temporal.
      Posterior communicating artery, 533. (524)
        Branches:—Uncinate.
          Middle thalmic.
      Anterior choroid, 533–4. (524)

The Veins of the Medulla and Pons, 663.   (649)
Medulla oblongata.
Pons.
The vessels and the remaining portion of the membranes should now be removed from the base of the brain. The superficial structures of the base and the superficial origin of each of the cranial nerves should then be noted.

Base of the Brain :—Dissection, 719-20.   (703-4)
The cerebral hemispheres, 720-1.   (704-5)
    Anterior pole, posterior.
    Fissures, or sulci.
    Convolutions, or gyri.
    Super-external surface.
    Internal surface.
    Inferior surface.
Fissures, 721-2-3.   (705-6-7)
    (*a*) Great longitudinal fissure.   Great transverse fissure.
    (*b*) Complete and incomplete.
    Interlobar fissures :—
        Parieto-occipital fissure :—
            Internal parieto-occipital or internal perpendicular fissure.
            External parieto-occipital fissure.
        Fissure of Sylvius :—
            Posterior limb.
            Ascending limb.
            Anterior limb.
            Fissure of Rolando, or central sulcus.
                Superior genu.   Inferior genu.
Lobes of the Cerebral Hemispheres with Fissures and Convolutions, 723-4.
(707-8)
    Frontal.   Parietal.   Occipital.   Temporo-sphenoidal.   Central lobe or
    Island of Reil.—Falciform lobe.   Olfactory lobe, rhinencephalon.

    Frontal lobe, 724-5.   (708-9)
        Surfaces :—Super-external or frontal.
            Inferior, or orbital.
            Internal, or mesial.
        Frontal surface, orbital margin.
            Præcentral sulcus.
            Superior frontal fissure.   Sulcus fronto-marginalis.
            Inferior frontal fissure.
            Ascending frontal convolution.
            Superior frontal convolution.
            Middle frontal convolution.
            Inferior frontal convolution :—
                Pars orbitalis.
                Pars triangularis.
                Pars basilaris.
        Orbital surface :—
            Tridate fissure.
            Olfactory or straight sulci.
            Orbital convolutions :—
                Internal.   Anterior.   Posterior.
            Straight convolution, or gyrus rectus.
Parietal lobe, sulcus subparietalis, 725-6.   (709-10)
    Intraparietal fissure.   Ramus occipitalis.
    Ascending parietal convolution.
    Superior parietal lobe.
    Supramarginal convolution.
    Angular gyrus.

Occipital lobe, 726–7–8. (710–11–12)
 Præoccipital notch.
 Superior occipital fissure. Ramus occipitalis.
  Transverse occipital fissure.
 Middle occipital fissure.
 Inferior occipital fissure.
 Superior occipital convolution.
 Middle occipital convolution.
 Inferior occipital convolution.
 Annectant gyri. Sulcus occipitalis anterior.
Temporal lobe, 728–29. (712–13)
 Parallel fissure.
 Middle temporal fissure.
 Inferior temporal fissure.
 Superior temporal or inframarginal convolution.
 Middle temporal convolution.
 Inferior temporal convolution.
 Transverse temporal convolutions.

Dissection.

Central lobe, or island of Reil. Fossa Sylvii. Limen insulæ, 729. (713)
 Sulcus circularis Reilii. Gyri operti. Sulcus centralis insulæ.
Mesial and tentorial surfaces of the hemisphere, 730–1–2–3. (713–14–15–16–17)
 Mesial surface. Internal occipital border.
  Tentorial surface.
Calloso-marginal fissure.
 Sulcus paracentralis.
 Sulcus subparietalis.
Parieto-occipital fissure.
Calcarine fissure.
Callosal fissure.
Dentate or hippocampal fissure.
Collateral fissure.
Marginal convolution.
Paracentral lobe.
Gyrus fornicatus, or gyrus cinguli. Isthmus.
Præcunius, or quadrate lobe.
Cuneus, or cuneate lobe.
Superior and inferior occipito-temporal convolutions.
Uncinate convolution. Uncus. Incisura temporalis. Tentorial groove.
 Substantia reticularis alba.
Lobus lingulis.
Inferior occipito-temporal convolution.
Impressio petrosa.
Falciform lobe ; limbic lobe.
 Outer segment. Inner segment.
 Olfactory tract, olfactory bulb, 788–9. (770–1)
  Cortex of the tract.
  External root. Middle root. Internal root.
Corpus callosum, 733–4–5. (717–18–19)
 Posterior extremity, or splenium.
 Anterior extremity, or genu.
 The rostrum. Basal white commissure.
 Body.

By successive slices remove the upper portion of the cerebral hemispheres down to the level of the corpus callosum, exposing its upper surface.

Striæ longitudinales.
Raphe.
Tæniæ tectæ.

Fibres of the corpus callosum : —

> Fibres from the genu. Forceps minor.
> Fibres from the body :—
> > Uppermost fibres.
> > Intermediate fibres.
> > Lowest fibres.
> > > Anterior set.
> > > Posterior set. Tapetum.
> > > Fibres from the splenium. Forceps major.

Expose the lateral ventricles, see Dissection, 735. (719)

The Lateral Ventricles, 735-6-7-8-9. (719-20-1-2)
> The body.
> Anterior cornu.
> Posterior cornu. Hippocampus minor, or calcar avis. Bulb of the cornu.

> Dissection.

Inferior or descending cornu.
> Floor. Roof.
> Eminentia collateralis.
> Hippocampus major, or cornu Ammonis.
> > Corpus fimbriatum, tænia hippocampi, or fimbria.
> Choroid plexuses.
> Foramina of Monro. Foramen commune anterius.
> Fascia dentata, or dentate convolution.
> Fascia cinearea.

> Dissection.

Basal ganglia of the Hemispheres, 739-40-1-2. (722-3-4-5)
> Nucleus caudatus :—
> > Head. Tail.
> Nucleus lenticularis.
> > Internal and external medullary laminæ.
> > > Globus pallidus.
> > > Putamen. Ansa lenticularis.
> Claustrum.
> Amygdaloid nucleus.
> Tænia semicircularis, or stria terminalis. Lamina cornea.
> Inner capsule, genu.
> Outer capsule.
> Anterior commissure.

> Dissection.

Fornix, 742-3. (725-6)
> Body. Posterior pillars. Lyre. Fimbria.
> > Anterior pillars.

Corpora albicantia. Bundles of Vicq d'Azyr.

Septum lucidum, 743. (726)
Fifth or Sylvian ventricle, 743. (726)

Great transverse fissure of the cerebrum, 743. (726-7)

Velum interpositum, or tela choroidea superior, 743-4.   727-8.
Choroid plexuses, 744. (728)

Dissection, 744-5. (728)

The Thalamencephalon, 745-6-7-8-9. (728-9-30-1-2-3)
> Third ventricle.
> > Posterior commissure.
> > Middle or gray commissure.
> > Floor.

> Central gray matter of the third ventricle.

Optic thalami.
    Anterior tubercle.
    Posterior tubercle, or pulvinar.
    Superior surface.
        Sulcus choroideus.
    Stria pinealis, or peduncle of the pineal body.
    Trigonum habenulæ.
    Internal surface.
        Middle or gray commissure.
    External surface.
    Inferior surface.

        Antero-superior, external and internal nuclei.
        Ansa lenticularis.

The pineal body.
    Stalk.
    Peduncle or stria pinealis.

        Recussus pinealis.

Posterior commissure.
Basal gray commissure.
Posterior perforated space.   Foramen cæcum anterius.
Corpora albicantia.   Bundle Vicq d'Azyr.
Tuber cinereum.   Infundibulum.   Recessus infundibuli.
Pituitary body, or hypophysis cerebri.

        Anterior lobe.   Posterior lobe.
        Pituitary gland.   Pituita.

Optic commissure.
Lamina cinearea.

Dissection.

The Mesencephalon, 749–50–1–2–3. (733–4–5–6)
    Aqueduct of Sylvius, or iter e tertio ad quartus ventriculorum.
        Lamina quadrigemina.   Corpora quadrigemina.

        Gray matter of the aqueduct.
            Nucleus of the third and fourth nerves.
            Nucleus of the descending root of the fifth nerve.

Corpora quadrigemina.
    Nates.   Testes.   Frenulum veli.
    Nates.   Brachia of the nates.

    Testes.   Brachia of the testes, or inferior brachia.
Corpus geniculatum internum.
Corpus geniculatum externum.
Crura cerebri :—
    (1) Tegmen.
    (2) Substantia nigra.
    (3) Crusta.
        Oculo-motor and lateral grooves.   Fillet.
    The crusta, or pes.

        Dissection.

        Crusta :—Central third.
            Outer third.
            Inner third.

Tegmental portions.   Subthalamic body.
    Tegmental or red nucleus.
    Superior cerebellar peduncle.
    Fillet.
    Posterior longitudinal bundle.

The Epencephalon, 753-760. (736-743)

Dissection.

Cerebellum :—
    Superior vermiform process.   Valeculecula.   Incisura marsupialis.
    Folia. Fissures.
    Peduncles :—
        Superior.
        Middle.
        Inferior.
    Great horizontal fissure.
    Superior surface of the cerebellum.
        Superior vermiform process.
        Sulcus cerebelli superior.
            Quadrate lobe.
                Anterior crescentic.
                Posterior crescentic portion.
            Posterior superior lobe.
        Superior vermiform process :—
            Lobus centralis. Alæ. Lingula.
            Monticulus cerebelli :—
                Culmen.
                Declive.
                Folium cacuminis.
    Inferior surface of the cerebellum.
        Lobes :—Posterior inferior.
            Slender.
      ·   Biventral.
            Tonsillar lobe.
            Flocculus.
    Inferior vermiform process :—
        Tuber valvulæ.
        Pyramid.
        Uvula. Furrowed band.
        Nodule, or laminated tubercle. Inferior medullary velum.

Dissection.

White matter of the cerebellum.
    Arbor vitæ :—
        Vertical branch.
        Horizontal branch.
Gray matter in the interior of the cerebellum.
    Nucleus emboliformis.
    Nucleus globosus.
    Nucleus fastigii.
Fourth ventricle.
    Roof :—Valve of Vieussens, or superior medullary velum.
        Inferior medullary vela, commissuræ ad flocculos.
        Choroid plexuses.
        Tela choroidea inferior.
        Obex.
        Ligulæ.
    Floor of the fourth ventricle.
        Sulcus longitudinalis medianus. Ventricle of Arantius.
        Striæ medullares or striæ acusticæ.
        Posterior moiety of the floor.
            Fovea inferior or posterior. Ala cinerea, trigonum vagi.
            Trigonum hypoglossi.
            Tuberculum acusticum.

Floor of the Fourth Ventricle :—
    Anterior moiety of the floor.
        Eminentia teres.   Fasiculus teres.
        Fovea superior.   Conductor sonorus.
        Locus cæruleus.   Substantia ferruginea.
Pons varolii.
    Dorsal surface.
    Ventral surface.
    Superior border.
    Inferior border.   Oblique fasciculus.
    Ventral region.
        Transverse fibres :—
            (a) Commissural.
            (b) Decussating fibres.
            (c) Fibres.
        Longitudinal fibres.
    Upper or tegmental region.
The Metencephalon, 760-766.   (743-749)
    Medulla oblongata, or bulb.
    Anterior surface.
    Posterior surface.   Ventricular part.
    Fissures :—Anterior fissure.
            Foramen cæcum posterius, or foramen cæcum of Vicq d'Azyr.
            Sulcus lateralis ventralis.   Postolivary sulcus.
        Posterior surface.
            Posterior fissure.
            Sulcus paramedianus dorsalis.
            Sulcus lateralis dorsalis.
    Anterior area.
        Pyramidal bodies.
        Funiculus anterior.
    Lateral area.   Lateral column.
        Gray tubercle of Rolando.
        Olivary body.   External arciform fibres.
    Posterior area of the medulla.
        Funiculus gracilis.   Clava.
        Funiculus cuneatus.   Cuneate tubercle.
        Restiform body.
    Dissection.
Internal structure of the medulla.
    Arrangement of the gray matter.
    Olivary nucleus.
    Nucleus of the clava and the nucleus cuneatus.
Cranio-cerebral Topography, 766 to 771.   (749 to 754)
Guide points :—
    Glabella.   Naso-frontal groove.   Orbital arch.   External angular process.
    Zygoma.
    External auditory meatus.   Mastoid process.   Superior nuchal line.
    External occipital protuberance, or inion.
    Parietal eminence.   Sutures.   Lambda.
Relation of the margin of the hemisphere to the cranial wall.
Orbital margin of the hemisphere.   Sylvian point.
Position of the principal fissures :—
        Fissure of Rolando.
        Naso-lambdoidal line.
        Fissure of Sylvius.
        External parieto-occipital fissure.
    Angular gyrus.
    Supramarginal convolution.

The Peripheral Nervous System, 786. (768-9)
In the brain laboratory, with the base of a brain, models, diagrams, and prepared sections at hand, the student should study the general description of each cranial nerve ; the branches and distribution will be demonstrated on the subject as the dissection proceeds.

First or olfactory nerves, 788-9. (770-1-2)
      Olfactory tract.   Olfactory bulb.
          Cortex of the tract.
          External root, middle, internal root.
Optic nerves, 790-1. (772-3)
      Optic chiasma, or commissure.
          Constitution of the optic chiasma.
              Uncrossed fibres.
              Crossed fibres.
              Commissural fibres.
      Optic tract.
          Internal root.
          External root.
      Optic radiation.   Gudden's Commissure.   Meynert's Commissure.
Third or oculo-motor nerve, 791-2. (774-5)
Fourth or trochlear nerve, 793. (775-6)
Fifth or trigeminal nerve, 794-5. (776-7)
      Portio major, ascending root.
      Portio minor, nucleus, descending root.
      Ascending root :—
          Accessory sensory nucleus.
          Cerebellar root.
      Descending root :—
          Motor nucleus.
      Gasserian ganglion.
Sixth or abducent nerve, 805-6. (787-8)
Seventh or facial nerve, 806. (788-9)
      Genu nervi facialis.
      Pars intermedia of Wrisberg.
      Geniculate ganglion.
Auditory nerve, 811-12-13. (792-3-4)
          Lateral root.
          Mesial root.
      Small-celled or chief nucleus.
      Large-celled or Deiter's nucleus.
      Accessory nucleus.
      Tuberculum acusticum.   Nucleus funiculi teretis.   Superior olive.
      Lateral root :—
          Conductor sonorus.
      Mesial root.
Glosso-pharyngeal nerve, 813. (794-5)
          Small-celled nucleus.
          Large-celled nucleus.
      Ascending root.
      Jugular ganglion.
      Petrous ganglion.
Pneumogastric, or vagus nerve, 815-16-17. (796-7)
Spinal accessory nerve, 819-20. (800-1)
Hypoglossal nerve, 820-1. (801-2)

Landmarks of the Back, 1170-1180. (1146-1156)
Median furrow.
Vertebral spines.
Thoracic spines.

Landmarks of the Back :—
Lumbar spines.
Muscles.
Trapezius.
Latissimus dorsi.
Triangle of Pettit.
Origin of spinal nerves.
Scapula, its muscles and arterial anastomoses.
Lumbar fascia.
Viscera, see table indicating the position of organs in cervical, thoracic, lumbar, and sacral regions.

Dissection of the Posterior Cervical Structures and the Back.

The dissector of the Upper Extremity should participate in the dissection of the back, until the dissection of the serratus magnus is completed.

Make a median incision from the occipital protuberance to the sacrum. Make a transverse incision from the spine of the seventh cervical vertebra to the end of the acromion process, also from the spine of the last dorsal vertebra upward and outward to the same point. Make an incision from the spine of the last lumbar vertebra to the crest of the ilium and along the crest. Turn the skin outward from the median line. The dissectors of the lower extremity will at the same time expose the sacral, coccygeal, and gluteal regions. The superficial lumbar, sacral, and coccygeal nerves should be noted by the dissectors of the back, also.

Superficial fascia, 302. (305)
Superficial nerves, Fig. 452 ; P. 826. (P. 807)
    Cervical nerves, 825. (806)
        Internal branches, 3d, 4th, 5th.
    Thoracic nerves, 827. (808)
        Internal branches—from upper six or seven nerves.
        External branches—from lower five or six nerves.
    Lumbar nerves, 827–8. (808)
        External branches of upper three nerves.
    Sacral and coccygeal nerves, 828. (808–9)
        External branches of the upper three nerves. (See internal branches.)
        Lower two sacral and·the coccygeal nerve.
The superficial arteries are derived from branches of the intercostal and lumbar arteries and accompany the superficial nerves :—
            From intercostal arteries :—
                Dorsal branch, 582 ; Fig. 362. (571–2)
                (ii) Muscular branch.
            From lumbar arteries :—
                Dorsal branch, 588. (577)

Turn off the superficial fascia and expose the deep fascia.

The deep fascia, 302. (305)

Reflect the deep fascia from the surface of the trapezius and latissimus dorsi.

Trapezius, 302–4 ; Fig. 262. (305–7)
    Origin. Insertion. Structure. Nerve-supply. Action. Relations. Variations.

Divide the trapezius near its origin and turn outward to the insertion. Note the spinal accessory nerve and the branches from the cervical plexus entering the deep surface of the muscle.

Spinal accessory nerve, 819–20. (800–1)
Branches to the trapezius, 831. (812)
    Subtrapezial plexus.
Latissimus dorsi, 307–8 ; Figs. 262 and 269. (310)
    Origin. Insertion. Structure. Nerve-supply. Action. Relations. Variations.

Divide the latissimus dorsi transversely on a line with the vertebral border of the scapula, reflect the inner portion to the origin, note the relations, then remove the inner portion of the muscle. In the neck, trace the superficial cervical artery, expose the origin of the omo-hyoid—476—and follow the supra-scapular artery—546 (536)—and nerve—834 (815)—to the point where they enter the supra-spinous fossa.

Superficial cervical artery, 548   (538)
Levator anguli scapulæ, 305 ; Figs. 263 and 266.   (307-8-9)
   Origin.   Insertion.   Structure.   Nerve-supply.   Action.   Relations.   Variations.
The rhomboidei, 305-6-7 ; Fig. 263.   (309-10)
   Rhomboideus minor :—
      Origin.   Insertion.   Structure.   Nerve-supply.
   Rhomboideus major :—
      Origin.   Insertion.   Structure.   Nerve-supply.   Variations.
   Action.   Relations of the two rhomboidei.

Divide the rhomboids at their origin and turn outward to the insertion.   Note the nerve to the rhomboids and follow the posterior scapular artery along the border of the scapula.

Nerve to the rhomboids, 834.   (815)
Posterior scapular artery, 547.   (538)

Expose the outer surface of the serratus magnus, then draw the scapula forward and outward and expose the inner surface.   Demonstrate the origin and the insertion.

Serratus magnus, 313-14-15, Fig. 267.   (316-17)
   Origin ; First part, Second part, Third part.
   Insertion ; First part, Second part, Third part.
   Structure.   Nerve-supply.   Action.   Relations.   Variations.

Divide the trunks of the brachial plexus and tie them to a small splinter to preserve their relative position.   Cut the axillary artery at its commencement.   Divide the muscles connecting the extremity with the trunk and remove it.   The remaining structures of the back will now be exposed.

Serratus posticus superior, 436-7-8 ; Fig. 305.   (432)
   Origin.   Insertion.   Structure.   Nerve-supply.   Action.   Relations.   Variations.
Serratus posticus inferior, 438 ; Fig. 305.   (434)
   Origin.   Insertion.   Structure.   Nerve-supply.   Action.   Relations.

Divide the serratus posticus superior at its origin, and turn it outward to the insertion.

Splenius : splenius capitis, splenius colli, 439-40 ; 305.   (435)
   Splenius Capitis : —
      Origin.   Insertion.   Structure.   Nerve-supply.   Action.
   Splenius Colli :—
      Origin.   Insertion.   Structure.   Nerve-supply.   Action.
      Relations of the splenius.   Variations.

Divide the splenius at its origin and throw it upward to the insertion.

The vertebral aponeurosis, 438 ; Fig. 305.   (434)
Posterior aponeurosis or lumbar fascia, 434.   (430)

Divide the vertebral aponeurosis and lumbar fascia by an incision about an inch from, and parallel with, the spinous processes ; reflect outward and expose the erector spinæ and its divisions.

Erector spinæ, 440-1 ; Fig. 306.   (436)
   Origin.   Insertion.   Structure.   Nerve-supply.   Action.   Relations.   ,
   Outer Division, 441-2-3.   (436-7-8)
      Ilio-costalis, or sacro-lumbalis :—
         Origin.   Insertion.
      Accessorius ad ilio-costalis or ad sacro-lumbalis.
         Origin.   Insertion.
      Cervicalis ascendens :—
         Origin.   Insertion.   Structure.   Nerve-supply.   Action.   Relations.
   Middle Division ; 443-4-5.   (438-9-40)
      Longissimus dorsi :—
         Origin.   Insertion ; externally, internally.
      Transversalis colli :—
         Origin.   Insertion.

Erector Spinæ—Middle Division :—
    Trachelo-mastoid :—
        Origin. Insertion.
      Structure of the middle division. Nerve-supply. Action. Relations.

  Inner division, 445. (440–1)
    Spinalis dorsi :—
        Origin. Insertion. Structure. Nerve-supply. Action. Relations.

    Variations of the erector spinæ and its divisions.

Divide the trachelo-mastoid about two inches below its insertion, reflect the lower portion of the muscle to the origin, then remove it; carefully raise the upper portion and expose the occipital artery in the second part of its course.

The occipital artery :—
    Second part of its course, 518–19–20. (511–12)
      Branches :—Mastoid branch.
           Princeps cervicis, superficial branch.
           Muscular branches.
Complexus, 446; Fig. 306. (441)
    Origin. Insertion. Structure. Nerve-supply. Action. Relations. Variations.

Divide the complexus transversely near the middle, carefully raise and reflect each portion, noting the nerves passing through it. Expose the deep branch of the princeps cervicis, and the posterior divisions of the spinal nerves.

Princeps cervicis, deep branch, 520. (511–12)
Deep cervical branch—from the superior intercostal, 551; Fig. 342. (541)
Deep cervical vein, 652. (639)
Posterior primary divisions of the spinal nerves, 825–6–7–8; Fig. 452. (806–7–8–9)
    Cervical nerves :—
      Internal branches. External branches.
      Posterior division of the second cervical nerve.
           Great occipital nerve.
    Thoracic nerves :—Internal branches.
           External branches.
    Lumbar nerves :—Internal branches.
           External branches—of the upper three nerves.
                of the fourth lumbar nerve.
                of the fifth lumbar nerve.
Arteries of the cervical, thoracic, and lumbar regions :—
    Cervical region :—
      Princeps cervicis, 520. (511–12)
      Deep cervical branch, 551. (541)
      Muscular branches, from the second portion of the vertebral, 542. (532)
    Thoracic region :—From intercostal arteries :—
             Dorsal branch, 582; Fig. 362. (571–2)
             (ii) Muscular branch.
    Lumbar region :—From lumbar arteries :—
             Dorsal branch, 588. (577)
The veins accompany the arteries and terminate in the intercostal or lumbar veins, in the thoracic or lumbar regions.

Demonstrate the middle layer of the lumbar fascia, 434. (430). Divide the erector spinæ at its origin and remove the muscle, together with its divisions, in order to expose the underlying muscles to advantage.

Semispinalis dorsi, 446–7–8; Fig. 308. (443)
    Origin. Insertion. Structure. Nerve-supply. Action. Relations.
Semispinalis colli, 448; Fig. 308. (443)
   ·Origin. Insertion. Structure. Nerve-supply. Action. Relations.
Multifidus spinæ, 448; Fig. 308. (443–4)
    Origin. Insertion. Structure. Nerve-supply. Action. Relations.

Follow the branches of the sacral and coccygeal nerves through the multifidus spinæ, cutting away the muscle to expose the nerves. The entire muscle should then be removed.

Sacral and coccygeal nerves, 828. (808-9)
    Internal branches.   External branches.
    Lower two sacral and the coccygeal nerve.
Rotatores spinæ, 449. (444)
    Origin. Insertion. Structure.   Nerve-supply.   Action.   Relations.
Interspinales, 449-50. (444)
    Origin. Insertion. Structure.   Nerve-supply.   Action.   Relations.
Intertransversales, 449-50. (445)
    Structure. Nerve-supply. Relations.
Levatores costarum, 421 ; Fig. 299. (418)
    Origin. Insertion. Structure.   Nerve-supply.   Action.   Relations.
    Ligamentum nuchæ, 1250. (1226)
Spinal veins, 667-8 ; Fig. 390. (653-4)
    Extraspinal and intraspinal.
        The extraspinal veins:—
            Anterior spinal plexus.
            Posterior spinal or dorsal spinal plexus.

In the suboccipital triangle trace the branches and expose the posterior primary division of the first cervical nerve.

Posterior primary division of the first cervical nerve, 826-7. (806-7)
The Suboccipital Muscles, 450-1 ; Fig. 308. (445-6)
    Rectus capitis posticus major :—
        Origin. Insertion. Structure.   Nerve-supply.   Action.   Relations.
    Rectus capitis posticus minor :—
        Origin. Insertion. Structure.   Nerve-supply.   Action.   Relations.
    Obliquus capitis inferior :—
        Origin. Insertion. Structure.   Nerve-supply.   Action.   Relations.
    Obliquus capitis superior :—
        Origin. Insertion. Structure.   Nerve-supply.   Action.   Relations.
    Rectus capitis lateralis, Fig. 316.
        Origin. Insertion. Structure.   Nerve-supply.   Action.   Relations.
        Variations.

Vertebral artery, 540; Fig. 342. (532)
    The second or vertebral portion.
Vertebral artery, 540-1 ; Fig. 342. (532).
    The third or occipital portion.
        In front.   Behind.
Branches of the occipital artery, 520 ; Fig. 335. (511-12)
    Communicating.

Clean away the muscles, exposing the laminæ of the vertebræ. Saw through the laminæ of the vertebræ close to the inner side of the articular processes. Divide the ligamenta subflava and remove the posterior wall of the spinal canal, exposing the cord inclosed in its membranes.

The spinal arteries are derived from:—
            Vertebral arteries ;
                Lateral spinal branches, 542. (532)
            Intercostal arteries ;
                Spinal branch, 582. (571)
            Lumbar arteries ;
                Dorsal branch, 588. (577)
The intraspinal veins, 668 ; Fig. 390. (654-5)
    Meningeal extra-medullary or meningo-rachidian veins :—
        Anterior longitudinal spinal veins (exposed later).
        Posterior longitudinal spinal veins.

The Spinal Cord, Spinal nerves, etc.

The spinal cord, 771. (754)
The dura mater, 772-3 ; Figs. 435-6. (755-6)
Open the dural sheath by a median incision, and expose the arachnoid.

The arachnoid, 773 ; Fig. 436. (756)
The subarachnoid tissue, 774. (757)

Remove the arachnoid from a small portion of the cord and expose the pia mater.

The pia mater, 774. (757)
Ligamentum denticulatum, 774 ; Fig. 435. (757)
External characteristics of the spinal cord, 774–6. (757–9)
　　　　Conus medullaris.　Cervical enlargement.
　　　　Lumbar enlargement.　Cauda equina.

Demonstrate the origin of one or two of the spinal nerves.　With a pair of strong pliers, cut away the bone and expose them in their passage through the intervertebral foramina ; expose also their termination into anterior and posterior primary divisions.

The spinal nerves, 822–3–5 ; Fig. 436–7. (803–4–5–6)
　　　　Anterior roots.
　　　　Posterior roots.
　　　　Course and direction.　Cauda equina.
　　　　Classification and number.
　　　　General distribution :—
　　　　　　　　Posterior primary divisions.
　　　　　　　　Anterior primary divisions.

Remove the spinal cord with its membranes.　Review the relations and structure of the membranes and demonstrate the origin of the spinal nerves in the manner indicated, see Dissection, 771–2. (754–5) When the cord is removed, the anterior longitudinal spinal veins can be demonstrated.

Membranes of the cord and ligamentum denticulatum, 772–3–4. (735–6–7)
External characteristics of the spinal cord, 774–6. (757–9)

Expose and demonstrate the fissures of the cord.

Fissures, 776; Fig. 438. (759)
　　　　Anterior longitudinal fissure.
　　　　Posterior longitudinal fissure.
　　　　Antero-lateral and postero-lateral fissures or grooves.
Columns :—
　　　　　　　　Anterior.
　　　　　　　　Lateral.
　　　　　　　　Posterior.

Make a number of transverse sections to exhibit the internal structure of the cord, Fig. 439.

Internal structure of the cord, 776–7–8. (759–60–1)

The student should read the chapter on The Deep Origin of the Spinal Nerves, pages 778–86. (761–68) The minute structure of the spinal cord and the deep origin of the spinal nerves can be studied to advantage only with specially prepared and mounted sections and with the aid of the microscope.　This should be done in the Histological, or Brain Laboratory.

THE UPPER EXTREMITY.

Landmarks of the Thoracic Region, Shoulder, and Axilla.
Superficial Anatomy of the Thorax, 1134 to 1140. (1111 to 1118)
Bony points :—
　　　　Sterno-clavicular joint.
　　　　Acromio-clavicular joint.
　　　　Ribs.
The mamma.
Structures found in an intercostal space.
Parts behind the manubrium.
Outline of the lungs :—
　　　　Their relation to the chest wall.　The pleura.
Outline of the heart :—
　　　　Its relation to the chest wall.
　　　　The valves.

Outline of the Heart—the Valves:—
    Pulmonary valves.
    Aortic valves.
    Auriculo-ventricular openings.
    Tricuspid valves.
    Mitral valves.
Relation of the vessels to the wall of the thorax. Aortic arch.
    Innominate artery. Left common carotid. Left subclavian artery.
    Innominate veins, left, right.
    Venæ cavæ :—
        Superior. Inferior vena cava.
    Internal mammary artery.
Shoulder and Axilla, 1180 to 1184. (1156 to 1160)
    Surface marks:—
        Clavicle. Acromion process.
        Great tuberosity and upper part of shaft of humerus.
        Coracoid process. Head of humerus. Clavicle.
        Bicipital groove.
Characteristic roundness of the shoulder. Deltoid.
Pectoralis major. Pectoralis minor. Coraco-brachialis.
Axilla.

Dissection of the External Thoracic Region and Axilla.

Make a median incision from the upper to the lower end of the sternum, and three transverse incisions : the first, from the sternum outward along the anterior border of the clavicle, and downward in the anterior median line of the arm; the second, from the lower end of the sternum outward and upward along the lower border of the pectoralis major to the arm, joining the first incision ; the third, from the lower end of the sternum outward and backward to the lower end of the scapula. Beginning at the median line, reflect the skin outward. Expose the superficial fascia of the pectoral region, axilla, and upper third of arm. Expose the superficial nerves and vessels of the thoracic wall. At the upper part of the thorax note the fibres of origin of the platysma.

The superficial fascia, 308. (311)
Platysma myoides, 452. (447)
    Origin. Structure.
Superficial nerves:—
    Descending branches—of cervical nerves, 831. (812)
        Suprasternal twigs.
        Supraclavicular nerves.
        Supra-acrominal branches.
Anterior cutaneous nerves of the thorax, 845. (826)
Lateral cutaneous nerves of the thorax, 845. (825)
    Anterior branches. Posterior branches.
The superficial arteries of the thoracic region are small branches from :—
    The internal mammary, 549-50; Fig. 344. (540)
        Perforating or anterior perforating branches.
        Second, third, and fourth perforating.
    The intercostal arteries, 583 ; Fig. 362. (572)
        Lateral cutaneous branches.
        Mammary glandular branches.
Lymphatics of the thorax, 695 ; Fig. 399. (680).
    Parietal lymphatic vessels, 695-6.
        Superficial parietal lymphatics.
        Lymphatics of the mammary gland.
        Superficial parietal glands. Epigastric gland.
If a female subject, remove the mammary gland and demonstrate its structure.

The mammary glands, 1108-9-10; Figs. 657-8. (1084-5-6-7)
        Female mamma :—Mammilla, or nipple. Areola.
        Secreting organ, structure of:—Adipose fossæ. Sinus or ampulla.
        Mammilla. Areola.
        Variations according to age and functional activity.

The Mammary Glands—
    Vessels and nerves:—
        Arteries. Veins. Lymphatics. Nerves.

      Development. Abnormalities. Nipples.

Dissect off the superficial fascia and expose the pectoral fascia, and deep fascia of the upper third of the arm.

Pectoral fascia, 308–9. (311)
Deep fascia—of arm, 321–2. (323–4)

Reflect the deep fascia from the surface of the pectoralis major by dissecting parallel with the course of its fibres. Expose also the anterior portion of the deltoid and remove the deep fascia from the upper third of the front of the arm.

Pectoralis major, 309–10–11; Fig. 264. (311–12–13)
    Origin. Insertion. Structure; clavicular, sterno-costal portion.
    Nerve-supply. Action. Relations. Variations.

In the interval between the pectoralis major and deltoid, exposed by separating the contiguous borders of the muscles, are the cephalic vein and the humeral branch of the acromio-thoracic artery.

Cephalic vein, 680–1; Fig. 396. (666)
Humeral branch—of the acromio-thoracic artery, 555. (545)

Carefully reflect the clavicular portion of the pectoralis major by detaching it from the clavicle and turning it downward; note the branches of the external anterior thoracic nerve entering its deep surface. The clavi-pectoral fascia should be carefully exposed in the interval between the clavicle and the pectoralis minor, and its connections and relations noted.

Clavi-pectoral fascia, costo-coracoid membrane, 308. (311)

Remove the costo-coracoid membrane and expose the subclavius muscle. Trace the cephalic vein to its termination and expose that portion of the axillary vein between the clavicle and the upper border of the pectoralis minor. Also, the first portion of the axillary artery and that part of the brachial plexus in relation with the vessels. Note the thoracic branches of the brachial plexus crossing in front of the vessels to the pectoral muscles. Demonstrate the thoracic axis and its branches and expose the origin of the superior thoracic artery.

Subclavius, 311–12–13; Fig. 265. (313–14)
    Origin. Insertion. Structure. Nerve-supply. Action. Relations. Variations.
Axillary vein, 682. (668)
The axillary artery, 552. (542)
    First part of the axillary artery:—
        Relations:—In front. Behind. Outer side. Inner side.
        Branches of the first part of the axillary artery, 555. (543–5)
        Acromio-thoracic or thoracic axis:—
            Acromial branch.
            Pectoral branch.
            Clavicular branch.
        Superior or short thoracic.
External anterior thoracic nerve, 835. (815)
Internal anterior thoracic nerve, 835. (815)

Turn off the superficial fascia of the axilla, exposing the deep fascia or axillary fascia.

Axillary fascia, 308–9. (311)

Beginning at the anterior border of the axilla, separate the deep fascia from the lower border of the pectoralis major and carefully reflect it backward to the posterior border of the space. In the reflection of the fascia care must be exercised not to injure the lateral cutaneous branches of the second and third intercostal nerves.

Intercosto-humeral nerve, 845; Fig. 460. (825)
Lateral cutaneous branch of the third intercostal, 845. (825–6)
Axillary glands, 695; Fig. 399. (679–80)

Carefully dissect out the lymphatics, fat, and connective tissue, exposing the third portion of the axillary artery and the divisions of the brachial plexus in relation with it.

Third part of the axillary artery, 553. (543)
    Relations:—In front. Behind. Outer side. Inner side.

Divide the pectoralis major near the middle, turn the outer portion to the insertion; reflect the inner portion to the origin. Preserve the arteries, so far as possible, in connection with the main trunks.

Pectoralis minor, 313; Fig. 266. (314-15-16)
    Origin. Insertion. Structure. Nerve-supply. Action. Relations. Variations.
Second part of the axillary artery, 552-3. (542-3)
    Relations :—In front. Behind. Inner side. Outer side.

Divide the pectoralis minor near the middle and turn to its origin and insertion. The subclavius should also be divided at its insertion and thrown in toward its origin. The axillary artery will be fully exposed throughout its course; its relations should be reviewed and the branches demonstrated.

    Relations of :—First part of axillary artery, 552. (542)
            Second part of axillary artery, 552-3. (542-3)
            Third part of axillary artery, 553. (543)
    Chief variations in the axillary artery, 553. (543)
    Branches of the first part of the axillary artery, 555.

        Superior or short thoracic.
        Acromio thoracic or thoracic axis :—
            Acromial branch. Humeral branch.
            Pectoral branch. Clavicular branch.

Branches of the second part of the axillary artery, 555. (545)
        Long thoracic artery.
        Alar thoracic.
Branches of the third part of the axillary artery, 555-6-7-8. (545-6-7-8)
        Subscapular artery :—
            Dorsal scapular.
            Infrascapular.

        Chief variations in the subscapular artery.

        Anterior circumflex :—

            Branches :—Bicipital.
                    Pectoral.
            Chief variations.

        Posterior circumflex :—
            Branches :—Nutrient.
                 Acromial.
                 Muscular.

        Chief variations in the posterior circumflex.
Axillary vein, 682. (668)

Remove the middle third of the clavicle. Dissect off the fascia and connective tissue and expose the nerves of origin, trunks, divisions, cords, and terminal branches of the brachial plexus.

The brachial plexus, 832-3-4; Fig. 455. (813-14)
    The brachial plexus is formed by the anterior primary divisions of :—
            Four lower cervical nerves.
            First thoracic nerve.
    Usually joined by communicating branches from :—
            Fourth cervical and second thoracic nerves.
    The brachial plexus is divisible into four stages :—
    First stage :—All the component nerves are separate from each other.
            Four lower cervical and first thoracic nerve.
    Second stage:—Nerves unite with one another to form trunks.
            Upper trunk. Middle trunk. Lower trunk.
    Third stage :—Trunks divided into three anterior and three posterior divisions.
    Fourth stage :—Three posterior divisions unite to form posterior cord.
            Anterior divisions of upper and middle trunks unite to form outer cord.
            Anterior division of inner trunk continued as inner cord.
    Position of :—The first stage of the plexus.
        Second stage. Third stage. Fourth stage.

Trace the branches of the brachial plexus through the region, or from their origin to the point where they pass into or under other structures. The terminal distribution of each branch will be demonstrated later, during the process of the dissection.

Branches of the brachial plexus:—
  Branches given off above the clavicle, 834–5. (814–15)
    Suprascapular nerve. Nerve to rhomboids.
    Posterior thoracic nerve. Communicating twig to phrenic.
    Nerve to subclavius.
  Branches given off below the clavicle:—
    Axillary or short branches, 835. (815–16)
      External anterior thoracic.
      Internal anterior thoracic.
      Subscapular nerves:—
        Upper or short subscapular nerve.
        Middle or long subscapular nerve.
        Lower subscapular nerve.
  Terminal or long branches: 835.
    From outer cord:—
      Musculo-cutaneous and outer head of median.
    From inner cord:—
      Inner head of median. Ulnar. Internal cutaneous
      and lesser internal cutaneous.
    From posterior cord:—
      Musculo-spiral and circumflex.

The dissector of the upper extremity should participate in the dissection of the back, until the dissection of the serratus magnus is completed. (See pages 58, 59)

The Shoulder.

Note the distribution of the cutaneous nerves, Fig. 456; P. 836. (817)

Remove the skin from the shoulder and the upper half of the arm, and expose the superficial fascia and the nerves.

Superficial fascia, 315. (318)
Superficial nerves:—
  Supra-acromial branches, 831. (812)
  Cutaneous branches—of the circumflex nerve, 835. (816)
    Anterior division. Posterior division.

Turn off the superficial fascia and expose the deep fascia.

The deep fascia, 315. (318)

Remove the deep fascia and expose the deltoid.

Deltoid muscle, 316–17; Figs. 262 and 264. (318–19)
  Origin. Insertion. Structure. Nerve-supply. Action. Relations. Variations.

Divide the deltoid at its origin and turn it down to the insertion. Note the vessels and nerves entering its deep surface.

Posterior circumflex artery, 557–8; Fig. 348. (547–8)
  Branches:—Nutrient. Articular. Acromial. Muscular.
    Chief variations in the posterior circumflex.

Circumflex nerve, 835. (816)
  Anterior division. Posterior division.
Anterior circumflex artery, 557. (546–7)
  Branches:—Pectoral. Bicipital.
    Chief variations.

Turn off the deep fascia covering the infra-spinatus, teres, and supra-spinatus.

Infra-spinatus, 317–18; Fig. 268. (319–20–1)
  Origin. Insertion. Structure. Nerve-supply. Action. Relations. Variations.
Teres minor, 319; Fig. 268. (321)
  Origin. Insertion. Structure. Nerve-supply. Action. Relations.

Teres major, 320–1 ; Figs. 268–9. (322–3)
    Origin. Insertion. Structure. Nerve-supply. Action. Relations.

Cut away the outer portion of the acromion process and expose the supra-spinatus.

Supra-spinatus, 317 ; Fig. 268. (319)
    Origin. Insertion. Structure. Nerve-supply. Action. Relations. Variations.

Subscapularis muscle, 319–20 ; Fig. 269. (321–2)
    Origin. Insertion. Structure. Nerve-supply. Action. Relations. Variations.

Carefully raise the supra-spinatus, infra-spinatus, and subscapularis from the scapula and expose the vessels and nerves passing under them.

Posterior scapular, 547–8 ; Fig. 343. (538)
    Branches:—Supraspinous. Infraspinous branches. Subscapular branches.
Suprascapular artery, 546 ; Fig. 343. (536–7)
    Branches of the suprascapular.
Suprascapular nerve, 834. (814–15)
Subscapular artery, 555–6–7 ; Fig. 343. (545–6)
    Branches:—Dorsal scapular. Infrascapular.

    Chief variations in the subscapular artery.

Landmarks of the Arm, Elbow, and Forearm.
    The Arm, 1186–7–8. (1162–3–4)
    The shaft of the humerus. Biceps. Brachial artery.
    Centre of the arm.
    Brachialis anticus.
    Median nerve.
    Ulnar nerve.
    Musculo-spiral nerve.
    The Elbow, 1188–9–90–1. (1164–5–6–7)
    Bony points.
    Hollow in front of the elbow.
    M-like arrangement of the superficial veins.
    The Forearm, 1193–4–5–6–7. (1168–9–70–1–2)
    Bony landmarks.
    Soft parts. Ulnar artery. Ulnar nerve. Median nerve.
            Radial artery. Radial nerve.
            Anterior interosseous artery.
            Posterior interosseous.

Dissection of the Arm and Forearm.
Outline the superficial nerve supply of the arm and forearm, Fig. 456 ; P. 836. (817)

In order to expose the superficial nerves and veins to the best advantage, remove the skin from the arm and reflect it from the forearm. To do this, continue the anterior median incision to the wrist, make a circular incision about two inches below the elbow and remove the skin from the arm above that point. Make a transverse incision at the wrist, reflect the skin of the forearm each way from the median incision to the posterior surface. On the posterior surface, beginning near the elbow, turn it toward the hand to the wrist. The skin of the forearm should not be removed, but preserved as a cover for the muscles after the nerves and veins are exposed, until the dissection of the arm is completed.

Superficial nerves of the arm and forearm, Fig. 456 ; P. 836. (817)
    Outer side :—
        Posterior division of the circumflex, 835.
        External cutaneous branches of the musculo-spiral, 841. (821–2)
                Upper branch. Lower branch.
        Musculo-cutaneous, 837–8. (816–17)
            Anterior branch. Posterior terminal branch.
    Inner side :—
        Intercosto-humeral, 845. (825)
        Internal cutaneous branch of musculo-spiral nerve, 841. (821)
        Lesser internal cutaneous nerve or nerve of Wrisberg, 835–6. (816)
        Internal cutaneous, 836.
            Anterior branch. Posterior branch.

Veins of the upper limb, superficial and deep, 678. (664)
    Superficial veins, 678; Fig. 396. (664-5)
        Superficial radial.
        The anterior and posterior superficial ulnar veins unite to form a single
        superficial ulnar vein.
        Superficial median vein :—
            Divides into :—
                Median cephalic—which unites with the superficial radial to form
                    the cephalic.
                Median basilic—which unites with the superficial ulnar to form the
                    basilic. (665)
                    The basilic joins with the inner brachial vena
                    comes to form the axillary vein, 679. (665)
    Superficial median vein, deep median vein, 679. (665)
        Median basilic vein, 680. (666)
            Anterior superficial ulnar, 679. (665)
            Posterior superficial ulnar, 679. (665)
                Basilic vein, 681-2. (666-7)
        Median cephalic vein, 679-80. (665)
            Superficial radial vein, 679. (665)
                Cephalic vein, 680-1. (666)
Lymphatics of the upper extremity, 693. (678)
    Superficial lymphatic vessels, 693; Fig. 399. (678)
    Superficial lymphatic glands, 693. (678)

Replace the skin of the forearm and bandage it in position until the dissection of the arm is completed.
Remove the superficial fascia from the arm and upper two inches of the forearm.

The deep fascia, 321-2. (323-4)

Divide the deep fascia in the median line and turn it out and in from the median incision, exposing the
biceps. The intermuscular septa should be demonstrated as the dissection proceeds. Push aside the
inner border of the biceps, dissect off the sheath and expose the brachial vessels in place.

The brachial artery, 559-60-1; Figs. 350-1. (548-9-50-1)
    Relations :—In front. Behind. External. Internal.
    Chief variations in the brachial artery.
    Branches, 561-2-3. (551-2-3)
        Superior profunda, (origin).
        Inferior profunda, chief variations.
        Anastomotica magna :—
            Branches:—Anterior. Posterior.
            Chief variations.
        Nutrient artery of the humerus.
        Muscular branches.
Deep veins of the upper extremity, 682. (668)
    Brachial venæ comites.
Deep lymphatic vessels, 693-4-5. (678)
Deep lymphatic glands, 695. (678)
Median nerve, 838. (817-18-19)
Internal cutaneous nerve, 836. (816)
Lesser internal cutaneous nerve, 835-6. (816)
Ulnar nerve, 840. (820)
Musculo-cutaneous nerve, 837. (816-17)
    Nerves to the coraco-brachialis.
    Nerves to the biceps and brachialis anticus muscles.
Coraco-brachialis, 321; Figs. 270-1. (323)
    Origin. Insertion. Structure. Nerve-supply. Action. Relations. Variations.
Biceps, 323-4; Fig. 270. (325-6)
    Origin, long head, short head. Structure. Nerve-supply.
    Action. Relations. Variations.

Brachialis anticus, 325 ; Fig. 270. (326-7)
    Origin. Insertion. Structure. Nerve-supply. Action. Relations. Variations.

Remove the deep fascia from the posterior surface of the arm and expose the triceps. Trace the musculo spiral nerve and the superior profunda artery.

Triceps extensor cubiti, 326-7 ; Fig. 268. (327-8-9)
    Origin, long head, external head, internal head. Insertion.
    Structure. Nerve-supply. Action. Relations. Variations.
Musculo-spiral nerve, 841. (821-2)
    Branches :—Internal cutaneous branch.
        External cutaneous branches :
            Upper branch. Lower branch.
        Nerve to the long head. Nerves to the outer and inner heads.
        Nerve to the anconeus.
        Nerves to the brachialis anticus, supinator longus and extensor carpi
            radialis longior.
Superior profunda artery, 562. (551-2)
    Branches:—Ascending branch. Cutaneous branch. Articular branch.
        Nutrient artery. Muscular branches.

Chief variations.

Forearm.
Remove the skin and superficial fascia to the wrist.

Deep fascia, 327-8. (329)
    Anterior annular ligament.

The deep fascia of the lower portion of the forearm, above the annular ligament, should be carefully reflected from the surface of the muscles. Care must be exercised not to remove the cutaneous nerve branches which pierce the deep fascia near the wrist and pass for distribution to the palmar and dorsal surfaces of the hand. At the upper part of the forearm near the internal condyle, the deep surface of the fascia gives origin to muscles and cannot be removed without injury to the muscles. The muscles should be carefully exposed and separated and the vessels and nerves displayed.

Brachio-radialis or supinator radii longus, 339 ; Fig. 272. (340)
    Origin. Insertion. Structure. Nerve-supply. Action. Relations. Variations.

That part of the brachio-radialis in relation with the radial artery should be raised and pushed aside to expose the vessel. The muscle will be considered later.

The radial artery, 571 ; Fig. 355. (561)
    The radial artery in the forearm, 571-2. (562-3)
        Relations :—In front. Behind. Outer side. Inner side.
    Variations in the radial artery in the forearm, 572-3. (563)
    Branches of the radial artery in the forearm, 573-4. (563)
        Radial recurrent. Muscular branches of the radial artery.
        Anterior radial carpal. Superficial volar.
Radial vein accompanies the artery.
Radial nerve, 841-2. (823)
Pronator radii teres, 328-9-30 ; Fig. 272. (330-1)
    Origin. Insertion. Structure. Nerve-supply. Action. Relations. Variations.
Flexor carpi radialis, 330-1 ; Fig. 272. (832)
    Origin. Insertion. Structure. Nerve-supply. Action. Relations. Variations.
Palmaris longus, 331 ; Fig. 272. (332)
    Origin. Insertion. Structure. Nerve-supply. Action. Relations. Variations.
Flexor carpi ulnaris, 321-2 ; Fig. 272. (333)
    Origin. Insertion. Structure. Nerve-supply. Action. Relations. Variations.
Flexor sublimis digitorum, 332-3-4-5 ; Fig. 273. (334-5-6)
    Origin. (Insertion.) Structure. Nerve-supply. Action. Relations. Variations.
The ulnar artery, 563-4. (553)
    Relations of the ulnar artery in the forearm, 564-5. (553-4)
        In front. Behind. Outer side. Inner side.
    Variations of the ulnar artery in the forearm. (555)

Branches of the ulnar artery in the forearm, 565-6-7-8.   (555-6-7)
   Anterior ulnar recurrent.   Posterior ulnar recurrent.
   Interosseous, or common interosseous artery.
   Muscular branches of the ulnar artery.
   Nutrient artery of the ulna.
   Posterior ulnar carpal.   Anterior ulnar carpal.
Ulnar vein accompanies the artery.
Ulnar nerve, 840.   (820-1)
   Branches:—Nerves to the flexor carpi ulnaris and flexor profundus digitorum.
      Origin of palmar cutaneous and dorsal or posterior branch.
Median nerve, 838.   (817-18-19)
   Branches:—Nerve to the pronator radii teres.
      Nerves to the flexor carpi radialis, palmaris longus, and flexor sub-
         limis digitorum.
      Origin of—palmar cutaneous branch.
Flexor profundus digitorum, 335-6-7 ; Fig. 275.   (336-7-8)
   Origin. (Insertion. Structure.) Nerve-supply. Action. Relations. Variations.
Flexor longus pollicis, 337-8 ; Fig. 275.   (338-9)
   Origin. (Insertion.) Structure. Nerve-supply. Action. Relations. Variations.
Pronator quadratus, 338 ; Fig. 284.   (339)
   Origin. Insertion. Structure. Nerve-supply. Action. Relations. Variations.
Anterior interosseous artery, 566.   (555-6-7)
   Branches of the anterior interosseous artery :—
      Arteria comes nirvi mediani, or median.
      Muscular branches.   Nutrient arteries.
      Anterior terminal, anterior carpal rete.
Anterior interosseous nerve, 838.   (819)

Landmarks of the Wrist and Hand, 1198 to 1204.   1173 to 1180.
   Bony points.
   Skin folds.
   Thenar and hypothenar eminences.
   Superficial palmar arch.   Deep palmar arch.

Dissection of the Wrist and Hand.
Continue the median incision to the interval between the second and third fingers. Make a transverse
incision at the base of the fingers, and median incisions to the end of each finger and the thumb. Turn
the skin each way from the median incisions.
Cutaneous nerves of the palm: Fig. 456. P. 836. (P. 817)
      Palmar cutaneous branch of the radial nerve, 841-2. (823)
      Palmar cutaneous branch—of the median nerve, 838. (819)
      Palmar cutaneous branch—of the ulnar nerve, 840. (821)
Anterior median plexus—of veins, 679. (665)
Palmaris brevis, 352-3-4 ; Fig. 280.
   Origin. Insertion. Structure. Nerve-supply. Action. Relations. Variations.
Palmar fascia, 351-2. 1206-7. (1182)
   Outer division, or thenar fascia.
   Inner division, or hypothenar fascia.

Divide the palmar fascia along the digital border of the annular ligament, leaving the ligament in place.
Turn the fascia forward toward the fingers, noting its relations to the deeper structures, then remove it.

The ulnar artery at the wrist, 568. (557-8)
   Relations :—In front. Below. Internally. Externally.
Superficial branch, superficial palmar arch, 568-9. (558-9-60)
   Relations:—In front. Behind.
   Variations in the superficial palmar arch, 569-70. (560)
   Branches of the superficial palmar arch, 570-1. (560)
      Digital arteries :—
         First digital artery.
         Second, third, and fourth digital arteries.
            Collateral digital arteries.

Branches of the Superficial Palmar Arch :—
  Muscular branches.
  Cutaneous branches.
  Deep branch.
Median nerve, 838–9 ; Fig. 458. (817–18–19)
  Branch to the muscles of the thumb.
  External terminal division, outer branch, inner branch.
  Internal terminal division, outer branch, inner branch.
  Pacinian corpuscles.
Ulnar nerve, 840 ; Fig. 458. (820–1)
  Superficial terminal branch :—
    Inner branch.   Outer branch.
Anterior annular ligament, 1204–5. (1180–1)
  Synovial membranes, synovial sacs, 1207–8–9.   (1182–3–4)
Ligamenta vaginialia, 352 ; Figs. 279–80.
Sheaths of the flexor tendons, or thecæ, 352.
  Vincula accessoria, ligamenta brevia, ligamenta longa.
  Theca of thumb.   Theca of little finger.

Divide the anterior annular ligament and expose the structures passing underneath.   Open the sheaths
of the tendons and trace each tendon to its terminal insertion.

Flexor sublimis digitorum, 332–3–4–5 ; Fig. 280.   (334–5–6)
  Insertion.   Structure.   Action.   Relations.   Variations.
Flexor longus pollicis, 337–8.   (338–9)
  Insertion.   Structure.   Action.   Relations.   Variations.

Cut the ulnar artery just below the deep or communicating branch and throw forward the superficial
palmar arch.   Divide the flexor sublimis digitorum and the median nerve at the middle of the forearm
and throw forward, exposing the flexor profundus digitorum and the lumbricales.

Flexor profundus digitorum, 335–6–7 ; Figs. 279–80–3.   (336–7–8)
  Insertion.   Structure.   Action.   Relations.   Variations.
Lumbricales, 354–5–6 ; Figs. 283 and 279.   (354–5)
  Origin.   Insertion.   Structure.   Nerve-supply.   Action.   Relations.   Variations.
Muscles of the thenar eminence, 358.
  Abductor pollicis, 358–9–60 ; Fig. 280.   (358–9)
    Origin.   Insertion.   Structure.   Nerve-supply.   Action.   Relations. Variations.

Divide the abductor pollicis near its origin and turn it forward to the insertion.

  Opponens pollicis, 360 ; Fig. 283.
    Origin.   Insertion.   Structure.   Nerve-supply.   Action.   Relations.
  Flexor brevis pollicis, 360–1–2 ; Figs. 283–4.   (360–1)
    Origin, outer head, inner head.   Insertion.   Structure.
    Nerve supply, outer head, inner head.   Action.   Relations.
  Adductor pollicis, or adductor pollicis transversus, 362 ; Fig. 284.   (361–2)
    Origin.   Insertion.   Structure.   Nerve-supply.   Action.   Relations.   Variations.
Muscles of the hypothenar eminence, 363.   (362)
  Abductor minimi digiti, 363 ; Fig. 283.   (362)
    Origin.   Insertion.   Structure.   Nerve-supply.   Action.   Relations.

Divide the abductor minimi digiti near its origin and turn forward to the insertion.

  Flexor brevis minimi digiti, 363 ; Fig. 283.
    Origin.   Insertion.   Structure.   Nerve-supply.   Action.   Relations.   Variations.

Divide the flexor brevis minimi digiti at its origin and throw forward to the insertion.

  Opponens minimi digiti, 364 ; Fig. 284.   (363).
    Origin.   Insertion.   Structure.   Nerve-supply.   Action.   Relations.   Variations.
Deep branch of the ulnar nerve, 840.   (821)
Radial artery in the palm, 576.   (565–67)
  Variations in the deep palmar arch.

Branches of the deep palmar arch, 576–7–8–9 ; Figs. 355 and 359. (567–8)
    Princeps pollicis. Radialis indicis.
    Palmar interosseous arteries.
    Recurrent branches.
    Posterior communicating or perforating.
Back of the forearm and hand.

To remove the skin from the back of the hand, make a median incision from the wrist to the interval between the second and third fingers. Make a transverse incision at the base of the fingers, and median dorsal incisions to the end of each finger and the thumb. Turn the skin each way from the median incisions and remove.

Cutaneous nerves, Fig. 459 ; P. 842. (822)
    Dorsal, or posterior branch of the ulnar nerve, 840. (821)
    Radial nerve, 841–2. (823)
    The dorsal surface of the thumb, and of the first second, and third fingers, receive branches from the terminal divisions of the median, 839. (819)
    Nerve supply of the dorsal integument of the hand, 842–3. (823)
Dorsal venous plexus, 679. (665)
    Digital veins.
Deep fascia, posterior annular ligament, 327–8. (329)
    Dorsal fascia, 351.

Reflect the deep fascia from the posterior surface of the forearm and hand. Preserve the posterior annular ligament.

Brachio-radialis, or supinator radii longus, 339 ; Figs. 272 and 276. (340)
    Origin. Insertion. Structure. Nerve-supply. Action. Relations. Variations.
Extensor carpi radialis longior, 341 ; Figs. 276 and 275. (340–1–2)
    Origin. Insertion. Structure. Nerve-supply. Action. Relations. Variations.
· Extensor carpi radialis brevior, 341–2 ; Fig. 276. (342)
    Origin. Insertion. Structure. Nerve-supply. Action. Relations. Variations.
Extensor communis digitorum, 342–3–4 ; Figs. 276–7. (343–4–5)
    Origin. Insertion. Structure. Nerve-supply. Action. Relations. Variations.
Extensor minimi digiti, 344–5 ; Figs. 276–7. ' (345)
    Origin. Insertion. Structure. Nerve-supply. Action. Relations. Variations.
Extensor carpi ulnaris, 345 ; Fig. 276. (345–6)
    Origin, first head, second head. Insertion. Structure.
    Nerve-supply. Action. Relations. Variations.
Anconeus, 345–6 ; Figs. 276–8. (346)
    Origin. Insertion. Structure. Nerve-supply. Action. Relations. Variations.

To expose the deeper layer of muscles detach the externus carpi ulnaris, externus minimi digiti, externus communicans digitorum, and externus carpus radicis brevior at their origin and throw forward. Note the vessels and nerves entering their deep surface.

Extensor ossis metacarpi pollicis, 348 ; Fig. 278. (348–9)
    Origin. Insertion. Structure. Nerve-supply. Action. Relations. Variations.
Extensor brevis pollicis, 349 ; Fig. 278.
    Origin. Insertion. Structure. Nerve-supply. Action. Relations. Variations.
Extensor longus pollicis, 349–50 ; Fig. 278. (350)
    Origin. Insertion. Structure. Nerve-supply. Action. Relations. Variations.
Extensor indicis, 350–1 ; Fig. 278.
    Origin. Insertion. Structure. Nerve-supply. Action. Relations. Variations.

Divide the anconeus at its origin and reflect to the insertion.
Expose the supinator brevis.

Supinator radii brevis, 346–7–8 ; Fig. 278. (346–7–8)
    Origin. Insertion. Structure. Nerve-supply. Action. Relations. Variations.
Posterior interosseous artery, 566–7 ; Fig. 353. (557)
    Interosseous recurrent, or posterior interosseous recurrent.
    Muscular branches. Articular branches.
Posterior terminal—branch of anterior interosseous artery, 566. (557)
                    Posterior carpal rete.
Posterior interosseous nerve, 841. (823)

Radial artery at the wrist, 574-5-6 ; Fig. 358. (563-4-5)
    Relations.
    Branches of the radial artery at the wrist :—
        Posterior radial carpal :—
            Second and third dorsal interosseous arteries.
            Dorsal digital branches.
        Metacarpal, or first dorsal interosseous artery :—
            Dorsal digital arteries.
        Dorsalis pollicis. ·
        Dorsalis indicis.

Open and demonstrate the compartments of the posterior annular ligament.

Posterior annular ligament, 1206. (1181-2)
Back of the wrist and hand, 1210. (1186)
Transverse ligament, 255 ; Fig. 236. (260)

Divide the transverse ligament, separate the metatarsal bones and expose the interossei muscles and perforating or communicating arteries.

The interossei, 356-7-8 ; Figs. 281-2.
    Three palmar interossei :—
        Origin. Insertion. Structure. Nerve-supply.
    Four dorsal interossei :—
        Origin. Insertion. Structure. Nerve-supply.
        Action of the interossei muscles. Relations. Variations.
Posterior, communicating, or perforating arteries, 579 ; Fig. 359. (568)

The muscles of the shoulder, arm, forearm, and hand should now be removed. On the forearm remove the supinator brevis last. The removal of the structures should be made a process of careful review, and as each muscle is removed, the student should note again the origin, insertion, structure,
actions, etc., especially of the deeper muscles, which are fully exposed as the superficial muscles are
removed. Demonstrate the arterial anastomoses about the elbow joint. The ligaments of the joints
should then be exposed and demonstrated, and the structure of the joints studied.

Arterial anastomoses about the elbow joint, 1191-2. (116-7-8)
Supinator radii brevis, 346-7. (346-7-8)
    Origin. Insertion. Action.

Articulations.
The various kinds of articulations, 187-8. (196-7)
    Synarthrosis :—
        True sutures.
        False sutures.
        Grooved sutures.
    Amphiarthrosis.
    Diarthrosis :—
        Arthrodia.
        Ginglymus.
            Ginglymo-arthrodial.
        Enarthrodia.
        Trochoides.
The various movements of joints, 188-9. (197-8)
    Gliding.
    Angular.
    Rotation.
    Circumduction.

Articulations of the Upper Extremity.
Scapulo-clavicular union, 230-1-2-3. (236-7-8-9)
    (*a*) Acromio-clavicular articulation.
    (*b*) Coraco-clavicular ligaments.
    (*c*) Proper scapular ligaments.
        Coraco-acromial and transverse.

The acromio-clavicular joint.

> Class :—*Diarthrosis.* Subdivision :—*Arthrodia.*
> Capsular ligament.
> Interarticular fibro-cartilage.
> Synovial membrane.

The coraco-clavicular union.

> Coraco-clavicular ligament :—
>> Conoid ligament.
>> Trapezoid ligament.
> Arterial supply.
> Nerve-supply.
> Movements.

The proper scapular ligaments.

> Coraco-acromial ligament :—
>> Anterior band.
>> Posterior band.
> Transverse, coracoid, or suprascapular ligament.
> Inferior transverse, or spino-glenoid ligament.

The shoulder joint, 234–5–6–7–8.   (239–40–1–2–3–4)

> Class :—*Diarthrosis.* Subdivision :—*Enarthrodia.*
> Capsular ligament.
> Gleno humeral bands of the capsule :—
>> Inner. Inferior. Superior, or gleno-humeral ligament.
> Coraco-humeral ligament.
> Glenoid ligament. Biceps tendon.
> Articular cartilage.
> Synovial membrane.
> Transverse humeral ligament.
> Muscles—in relation—with the capsular ligament.
> Arterial supply. Nerve-supply.
> Movements.

The elbow joint, 239–40–1.   (244–5–6–7)

> Class :—*Diarthrosis.* Subdivision :—*Ginglymus.*
> Capsule :—
>> Anterior portion.
>> Posterior portion.
>> Internal portion.
>> External portion.
> Synovial membrane.
> Arterial supply.
> Nerve-supply.
> Movements.

The radio-carpal articulation, 245–6–7–8–9.   (251–2–3–4)

> Class :—*Diarthrosis.* Subdivision :—*Ginglymus.*
> Anterior radio-carpal.
> Posterior radio-carpal ligament.
> Internal lateral ligament.
> External lateral ligament.
> Synovial membrane.
> Arterial supply.
> Nerve-supply.
> Movements.

The union of the radius with the ulna, 241–2–3–4–5.   (247–8–9–50)
Superior radio-ulnar.
The union of the shafts,—the mid radio-ulnar union.
Inferior radio-ulnar.
The superior radio-ulnar joint.

> Class :—*Diarthrosis.* Subdivision :—*Trochoides.*
> Orbicular ligament.
> Synovial membrane.
> Arterial and nerve-supply.

The mid radio-ulnar union.

> Oblique ligament.
> Interosseous membrane. Inferior oblique ligament. Oblique ligament.

The inferior radio-ulnar joint.

> Class:—*Diarthrosis.* Subdivision:—*Lateral Ginglymus.*
> Anterior radio-ulnar ligament.
> Posterior radio-ulnar ligament. Interosseous membrane.
> Triangular fibro-cartilage.
> Synovial membrane.
> Arterial supply.
> Nerve-supply.

The movements of the radius.

The carpal joints, 249–50–1–2–3. (254–5–6–7–8)
> Joints of the first row.
> Joints of the second row.
> Medio-carpal.

The union of the first row of carpal bones.

> Class ;—*Diarthrosis.* Subdivision:—*Arthrodia.*
> Two dorsal ligaments.
> Two palmar ligaments.
> Two interosseous fibro-cartilages.
> Synovial membrane.

The union of the pisiform bone with the rest of the first row.

> Capsule.

The union of the second row of carpal bones.

> Class:—*Diarthrosis.* Subdivision:—*Arthrodia.*
> Three dorsal ligaments.
> Three palmar ligaments.
> Two interosseous ligaments.
> Synovial membrane.

The medio-carpal joint, or the union of the two rows of the carpus with one another.

> Ligaments:—Anterior or palmar medio-carpal.
>   Posterior, or dorsal medio-carpal ligament.
>   Transverse dorsal ligament.
> Synovial membrane.
> Arterial supply.
> Nerve-supply.
> Movements.

The carpo-metacarpal joints, 253–4–5. (258–9)
> Carpo-metacarpal joints of the four inner fingers.
> Carpo-metacarpal joint of the thumb.

The four inner carpo-metacarpal joints.

> Class:—*Diarthrosis.* Subdivision:—*Arthrodia.*
> The dorsal ligaments.
> The palmar ligaments.
> Interosseous ligament.
> Synovial membrane.
> Arteries.
> Nerves.
> Movements.

The carpo-metacarpal joint of the thumb.

> Class:—*Diarthrosis.* Subdivision:—*Arthrodia.*
> Capsular ligament.
> Synovial membrane.
> Arteries.
> Nerves.
> Movements.

The intermetacarpal articulations, 255. (260)

> Class:—*Diarthrosis.* Subdivision:—*Arthrodia.*
> Dorsal ligaments.
> Palmar ligaments.
> Interosseous ligaments.
> Arteries.
> Nerves.
> Synovial membrane.

The union of the heads of the metacarpal bones.

Transverse ligament.

The metacarpo-phalangeal joints, 256-7-8. (260-1-2)
   The metacarpo-phalangeal joints of the four inner fingers.

Class:—*Diarthrosis.* Subdivision:—*Ginglymus.*
Glenoid ligament.
Lateral ligaments.
Posteriorly.
Synovial sac.
Arteries.
Nerves.
Movements.

The metacarpo-phalangeal joint of the thumb.

Class:—*Diarthrosis.* Subdivision:—*Ginglymus.*
Lateral ligaments.
Posterior ligament.
. Sesamoid bones.
Arteries and nerves.
Movements.

The interphalangeal articulations, 258. (262-3)

Class:—*Diarthrosis.*' Subdivision:—*Ginglymus.*
Glenoid ligament.
Lateral ligaments.
Posteriorly.
Synovial membrane.
Arteries and nerves.
Movements.

THE THORAX AND THORACIC VISCERA.

The surface of the thorax should now be cleaned of all muscle tissues, exposing the ribs and external intercostal muscles.

The thorax, or trunk of the body, 935-6. (914-15)
   Formed :—In front.
         Behind.
         Laterally.
   Superior aperture :—
         Bounded :—In front.
               Behind.
               On each side.
      Structures passing through the upper aperture of the thorax, 1136. (1112-13)
            In the middle line.
            On each side.
      Lower opening, or base.

The structures of the thoracic wall should now be exposed in place, in the order and by the methods indicated.

Intercostal muscles, 417. (414)
   External intercostals, 417-18-19, (414)
      Origin. Insertion. Structure. Nerve-supply. Relations.
      External intercostal fascia.

Divide the external intercostal muscles and the external intercostal fascia near the lower border of each intercostal space. Reflect the external intercostal muscles upward and expose the internal intercostal muscles and the intercostal nerves and vessels.

Internal intercostals, 419-20-1. (416-17)
      Origin. Insertion. Structure. Nerve-supply. Relations.
      Internal intercostal fascia.
      The action of the intercostal muscles.

Expose and trace two or three of the intercostal nerves, as the second, sixth, and tenth. Exercise care not to injure the pleura.

Intercostal nerves, 843–5. (823–5–6)
    Upper group. Lower group.
    First thoracic nerve.
    Upper intercostal nerves.
        Lateral cutaneous nerves of the thorax.
        Anterior cutaneous nerves of the thorax.
    Lower intercostal nerves.
    Branches.

Expose and trace the intercostal arteries in two or three of the intercostal spaces, as the third, fifth, and seventh. In the anterior portion of the upper intercostal spaces, expose the internal mammary and anterior intercostal arteries.

Superior intercostal artery, 550. (540–1)
Aortic intercostal arteries, subcostal artery, 581–2. (570–1)
    The intercostal portion.
    Collateral intercostal branch.
Internal mammary artery, 548–9–50. (538–9–40)
    Thoracic portion :—
        Anterior intercostal arteries.
        Perforating, or anterior perforating branches.
Intercostal veins, subcostal, 644–5. (632)
Internal mammary veins, 640. (628)
Intercostal lymphatics, 696. (681)
Internal mammary, sternal, or anterior intercostal glands, 696. (681)

Remove the intercostal muscles from the anterior portion of the upper five intercostal spaces, exposing the outer surface of the parietal pleura. With the finger carefully separate the pleura from the anterior portion of the second, third, fourth, and fifth ribs and their costal cartilages, on each side; also from the intervening portion of the sternum. Divide the second, third, fourth, and fifth ribs at the axillary line. Divide the sternum transversely below the cartilages of the first ribs, also just above the cartilages of the sixth ribs. Beginning above, raise the sternum with its attached cartilages and ribs, drawing it forward and downward, carefully separating the pleura and internal mammary arteries from it. With a chain hook retain it at a right angle with the thoracic wall while the triangularis sterni is exposed, then remove it.

Triangularis sterni, 421–2; Fig. 301. (418–19)
    Origin. Insertion. Structure. Nerve-supply. Action. Relations.

Divide the pleura by a vertical incision about six inches from the median line, and by incisions parallel with the second and sixth ribs, to within about three inches of the median line. The flap thus formed can be turned forward, and the reflection of the pleura can be readily demonstrated, also the interval of non-approximation of the pleural sacs. The visceral and parietal portions of the pleura should then be traced and the relations studied.

The pleuræ, 958–60. (936–8–9)
    Visceral layer, or pleura pulmonalis.
    Parietal layer, or pleura costalis.
    Ligamentum latum pulmonalis.
    Right pleural sac.
    Inner surface.
    Outer surface.
Viscera within the thoracic cavity, 936. (915)
    Heart. Lungs.
Mediastinal space, 936–7–8. (915–16–17)
        Superior mediastinum.
        Anterior. Posterior.
        Middle mediastinum.
    Superior mediastinum :—
        Boundaries :—In front.
            Behind.
            On each side.
            Above.
            Below.
    Middle mediastinum.
    Anterior mediastinum.
    Posterior mediastinum.

The lungs, 958–60.   (936–7–8)

With the blowpipe inserted in the trachea, inflate the lungs and note their relations to the heart, thoracic walls, etc.

Outer surface.
Inner surface.
Base.
Apex.
Two lobes.   Third or middle lobe.

The branches of the internal mammary artery not already shown should now be exposed and traced; then divide the mammary vessels and throw them upward.

Internal mammary artery, 548–9–50.   (538–9–40)
Lateral infracostal artery.
Superior phrenic or comes nervi phrenici.
Mediastinal or thymic branches.
Pericardiac branches.
Sternal branches.
Superior epigastric artery.
Musculo-phrenic artery :—
Phrenic.
Anterior intercostals.
Muscular.

Strip the pleura from the superior mediastinum and pericardium, also from the adjacent upper surface of the diaphragm, exposing the relations of the pericardium to the diaphragm.   Expose the mediastinal glands.   Trace the phrenic nerves; then expose the pericardium.

Anterior mediastinal or sternal glands, 697.   (682)
Superior mediastinal or cardiac glands, 697.   (682)
Phrenic nerve, 832.   (812–13)
Right side.
Left side.
Branches.

The pericardium, 962–3.   (941–2)
Fibrous layer.   Ductus arteriosus.   Sterno-pericardial bands.

Make a transverse incision through the pericardium, extending between the roots of the lungs, also a median incision from the aorta to the apex of the heart.   Study the relations of the peri- cardium to the heart and the blood-vessels at its base; also the position and relations of the heart.

Serous layer :—
Parietal portion.
Visceral portion.
Oblique sinus.
Vestiginal fold of the pericardium.   Oblique vein of Marshall.
Relations.
On opening the pericardium the following structures may be observed :—
Right and left ventricles.   Right auricle.   Right auricular appendix.
Aorta.   Superior vena cava.   Pulmonary artery.   Left auricular appendix.
Left ventricle.   Coronary sinus.   Cardiac veins.   Oblique vein of Marshall.
Coronary arteries.
Vessels.
Lymphatics of the pericardium, 697.   (681–2)
The heart, 963–4–5.   (942–3)
Position :—Base.
Apex.
Lower border.
Lateral borders.
Size and weight.
Anterior surface.
Posterior surface.
Auriculo-ventricular groove.
Interventricular groove.

Expose the remains of the thymus gland, then clear it away. Demonstrate the superficial cardiac plexus and trace the superior and inferior cervical cardiac nerves of the left side. Carefully dissect away the pericardium, fat, connective tissue and fascia from the base of the heart. Expose the superior vena cava and its tributaries, also the arch of the aorta and its branches.

The thymus body or gland, 955-6-7. (935-6)
    Relations.
    Structure.
    Vessels :—Arteries.
            Veins.
    Nerves.
    Lymphatics of the thymus, 697. (682)
Superficial cardiac plexus, 869-70. (849)
    Superior cervical cardiac nerve, left side, 867. (847)
    Inferior cervical cardiac nerve of the left side, 819. (800)
The veins, 638. (626)
    Pulmonary.
    Systemic.
    Veins of the thorax.
Innominate, or brachio-cephalic veins, 640-1. (627-8)
    Right innominate vein :—
        Relations :—In front.
            Behind.
            Right.
            Left.
    Left innominate vein :—
        Relations :—In front.
            Behind.
            Below.
            Above.
    Tributaries :—
      Internal mammary veins.
      Left superior intercostal vein.
      Mediastinal, pericardial, and thymic veins.
Superior or descending vena cava, 639. (627)
    Relations :—In front.
          Behind.
          Right side.
          Left side.
    Tributaries.
    Chief variations in the superior vena cava and innominate veins, 642. (628-9-30)
        (1) Variations due to the persistence of the left duct of Cuvier.
        (2) Variations due to persistence of the left and suppression of the right duct
          of Cuvier.
The aorta, 495 to 501. (487 to 494)
    Arch of the aorta.   Thoracic aorta.
    Arch of the aorta :—
        Ascending portion of the arch :—
            Relations:—In front.
                Behind.
                Right side.
                Left side.
        Transverse portion of the arch :—
            Relations :—In front.
                Behind.
                Above.
                Below.
        Descending portion of the arch :—
            Relations:—In front.
                Behind.

Descending Portion of the Arch—Relations:—
>> Right side.
>> Left side.

Variations in the arch of the aorta :—
>> *Variations in the aortic arch itself :—*
>>> (*a*) *Variations depending on abnormalities in development of the ventral aortic stem.*
>>> (*b*) *Variations depending on abnormalities in development of the fourth and fifth aortic or branchial arches.*
>>> (*c*) *Variations depending on abnormalities in development of the dorsal aortic stems.*

Branches of the arch of the aorta :—
> Ascending portion.
> Transverse portion.
> Descending portion.
> Transverse portion, II, 503. (495)
>> Innominate.
>> Left common carotid.
>> Left subclavian artery.
>> Innominate, or brachio-cephalic artery, 503–4. (495–6)
>>> Relations :—In front.
>>>> Behind.
>>>> Right side.
>>>> Left side.

>> Variations in the innominate artery.

>> Branches of the innominate artery :—
>>> (1) Right common carotid.
>>> (2) Right subclavian.
> Thoracic portion of the left common carotid, 505–6. (497)
>> Relations :—In front.
>>> Behind.
>>> Right side.
>>> Left side.
> Left subclavian artery, 536–7. (527–8)
>> Relations :—In front.
>>> Behind.
>>> Right side.
>>> Left side.

The chief variations in the three primary branches of the transverse portions of the aortic arch, 501. (493)
>> (*1*), (*2*), (*3*), (*4*), (*5*), (*6*).

Strip the pleura from the anterior and posterior surfaces of the root of the lung. Trace the pulmonary artery and its divisions to the lung, and the pulmonary veins from the lung to the left auricle. Dissect out the areolar tissue, glands, etc., and expose and separate the component parts of the root of the lung. Special care should be exercised not to injure the nerve plexuses and the branches to and from them. The anterior pulmonary plexus consists of a few small fibres in front of the root of the lung ; the posterior plexus is larger and readily exposed. The pneumogastric nerves should be exposed on each side and traced to the posterior plexus. To expose the pneumogastric nerve and the posterior plexus on each side, draw the lung forward and over to the opposite side and retain it in place with hooks while exposing the structures behind the root.

Anterior pulmonary branches (plexus), 819. (800)
Pulmonary artery, conus arteriosus, 491–2–3–5. (483–4–5–7)
> Trunk of the pulmonary artery :—
>> Ductus arteriosus, or ductus Botalli.
>> Relations :—In front.
>>> Behind.
>>> Right.
>>> Left.

Right pulmonary artery :—
    Relations :—In its course to the lung :—
            In front.
            Behind.
            Above.
            Below.
        At the root of the lung :—
            Above and behind.
            Below and in front.
Left pulmonary artery :—
    Relations at the root of the lung :—
        Behind and below.
        Below and in front.
Pulmonary veins, 639.   (626–7)
Root of the lung, 960–1.   (939–40)
        Pulmonary artery, pulmonary vein, bronchial tube, bronchial vessels,
        anterior and posterior pulmonary plexuses, lymphatic glands, areolar
        tissue—all encased in a pleural covering.
    Right root.
    Left root.
    Arrangement of structures within the root :—
        From above downward :—
            Right side.
            Left side.
        From before backward.
Right vagus nerve, 817.   (798)
Left vagus nerve, 817.   (798)
    Inferior or recurrent laryngeal nerve, 818–19.   (799–800)
        Of the right side.
        Left side.
    Posterior pulmonary plexus, 819.   (800)

Divide the superior vena cava just below the termination of the vena azygos major (642; Fig. 382).
Divide the right pulmonary artery near the right lung; draw the arch of the aorta and the base of the
heart forward and to the left and expose the deep cardiac plexus, tracing the branches to and from it.
The thoracic portion of the trachea and the bronchial tubes will also be exposed.

Cardiac plexus, 869–70.   (849–50)
    Deep cardiac plexus.
Cardiac branches—of the pneumogastric and sympathetic nerves, 819, 867.   (800, 847)
    Superior cervical cardiac nerve.
    Middle cardiac nerve.
    Inferior cervical cardiac nerve, right.
    Thoracic cardiac branches.
Right coronary plexus, 870.   (850)
Left coronary plexus, 870.   (850)
Trachea or air tube, 950–1.   (929)
    Thoracic portion.
The bronchi, 952.   (931)
    Right bronchus :—
        Relations.
    Left bronchus :—
        Relations.
Bronchial glands, broncho-mediastinal trunk, 697–8.   (682)

Expose the œsophagus and trace the pneumogastric nerves to the diaphragm.

Œsophagus, 987–8.   (966–7)
    In the thorax :—
        Relations :—In front.
            Behind.
            Laterally.

Lymphatics of the thoracic portion of the œsophagus, 697. (682)
Right vagus nerve, 817. (798)
Left vagus nerve, 817. (798)

Expose the thoracic aorta and trace the visceral branches.

Thoracic aorta, 579–81. (568–70.)
    Relations:—In front.
             Behind.
             Right side.
             Left side.
    Branches of the thoracic aorta :—
             Visceral.   Parietal.
             Visceral branches :—
                 Pericardial.
                 Bronchial arteries :—
                     Right bronchial.
                     Upper left bronchial.
                     Lower left bronchial.
                     Branches to the bronchial glands and œsophagus.
                 Œsophageal arteries.
Bronchial veins, 645. (633)
Œsophageal veins, 645. (633)
Posterior mediastinal glands, 697. (682)
Thoracic duct, 698–9. (683–4)
    Thoracic portion of the thoracic duct.
        Relations :—In posterior mediastinum.
                 Superior mediastinum.
    Cervical portion of the thoracic duct.

The chief variations in the thoracic duct.

Thoracic portion of the gangliated cord, 868. (848)
    Branches :—
        External branches.
        Internal branches of the upper ganglia.
        Internal branches of the lower ganglia.
            Internal branches—upper series.
            Internal branches—lower series.
                Great splanchnic nerve.
                Lesser splanchnic nerve.
                Smallest splanchnic nerve.
Azygos veins, 642–4. (630–1–2)
    Vena azygos major :—
        In the posterior mediastinum :—
            Relations:—Left side.
                 In front.
                 Right side.
        As it curls over the root of the lung.
        Tributaries.
    Vena azygos minor, etc. :—
        Tributaries.
    Vena azygos tertia :—
        Tributaries.
The intercostal veins, 644–5. (632–3)
    On the right side:—
        Lower right superior intercostal vein.
    Left side :—
        Upper left superior intercostal vein.
        Lower left superior intercostal vein.
Aortic intercostal arteries, 581–2. (570–1)

Vertebral portion :—
    Right side.
    Left side.
Intercostal portion :—
    Pleural branches, (572).
Diaphragmatic branches, 583. (573)
Aberrans artery, 583-4. (573)
Intercostal nerves, 843-5. (825-6)
    Upper intercostal nerves.
    Lower intercostal nerves.
Intercostal or posterior intercostal lymphatic glands, 696. (681)
Infracostales, or subcostals, 422-3. (419)
    Origin. Insertion. Structure. Nerve-supply. Action. Relations. Endothoracic fascia.

Open the cavities of the heart in the order and by the methods indicated, exposing the structures within. The student should at the same time dissect the heart of an ox or the heart of a sheep, and demonstrate the action of the valves in the manner indicated.

The heart :—
    Four cavities, 965-6. (942-3)
        Venous side.
        Arterial side.
    The right auricle, 966-7-8. (943-4-5)

Make an incision through the auricular wall between the superior and inferior venæ cavæ, and from the middle of this incision to the end of the appendix, exposing the interior of the auricle.

    Endocardium.
    Sinus venosus, or atrium. Auricular appendix.
    Openings :—
        Superior vena cava.
        Inferior vena cava.
        Auriculo-ventricular opening.
        Coronary sinus.
        Foramina Thebesii.
    Superior caval opening.
    Inferior caval opening. Eustachian valve.
    Coronary sinus. Valve of Thebesius.
    The foramina Thebesii, vena Galeni.
    Cavity of the right auricle.
    Fossa ovalis. Annulus ovalis.
    Tubercle of Lower.

Introduce the end of the blowpipe into the right ventricle by passing it through the pulmonary artery and between the pulmonary semilunar valves. Place a ligature around the commencement of the pulmonary artery, then inflate the ventricle. The action of the tricuspid valve and the closing of the auriculo-ventricular opening can be readily demonstrated.

Open the right ventricle by two incisions as follows: Make a vertical incision through the ventricular wall, about a quarter of an inch to the right of the anterior interventricular groove and parallel with it, beginning above at the conus arteriosus and passing downward and around the end of the ventricle ; make a transverse incision through the upper part of the ventricular wall, extending from the upper end of the first incision to the right border of the ventricle, half an inch below and parallel with the auriculo-ventricular groove. Exercise care to divide the ventricular wall only, and not to injure the structures within.

Right ventricle, 968-9. (945-6-7)
    Form. Conus arteriosus.
    Inner surface, or body ; ridges, bands, columns :—
        Columnæ cerneæ ; musculi papillares, moderator band.
        Musculi papillares :—
            Anterior.
            Right.
            Posterior.

Auriculo-ventricular opening. Tricuspid valve, 970. (948-9)

Cut away the upper portion of the conus arteriosus enough to expose the ventricular surface of the semilunar valves, at the commencement of the pulmonary artery.   Insert the blowpipe in the end of the pulmonary artery, retaining it in place with a ligature.   Inflate the artery and demonstrate the action of the valves.

Open the pulmonary artery by a longitudinal incision and expose the pulmonary semilunar valves.

Orifice of the pulmonary artery, 969.   (937-8)
    Pulmonary semilunar valves :—
        Sinuses of Valsalva.
        Corpus Arantii.
        Lunulæ.

Open the left auricle by an incision through the auricular wall, extending from the right to the left pulmonary vein (upper), and a second incision from the middle of the first incision to the tip of the appendix.

Left auricle. Cavity, 970-1. (949-50)
    Openings :—Pulmonary veins.
        Auriculo-ventricular opening.
        Foramina Thebesii.

Pass the blowpipe through the aorta into the left ventricle, in the same manner as on the right side, and demonstrate the action of the bicuspid valve and the closing of the auriculo-ventricular opening.

Open the left ventricle by an incision through the ventricular wall, about a half an inch to the left of the interventricular grooves, beginning below the auriculo-ventricular groove and passing downward to the apex of the ventricle and upward along the opposite margin of the cavity.

Left ventricle, 971. (950)
    Columnæ carneæ.
    Musculi papillares.
        Chordæ tendineæ.
Auriculo-ventricular opening, 972. (950)
    Bicuspid valve.
    Os cordis.

Demonstrate the action of the aortic semilunar valves by inflating the aorta.

Orifice of the aorta, 971-2. (950)
    Semilunar valves.

Open the aorta by a longitudinal incision and expose the aortic semilunar valves.   Trace the vessels of the heart, laying them open with the point of the knife.

Coronary arteries, 502-3, 972-3.   (494-5, 951-2)
    Right coronary artery :—
        Right auricular branch.
        Preventricular branch.
        Right marginal branch.
        Posterior interventricular branch.
        Transverse branch.
    'Left coronary artery :—
        Left auricular branch.
        Large anterior interventricular branch.
        Left marginal branch.
        Terminal branch.
    Variations in the Coronary Arteries.
Cardiac or coronary veins, 646-7-8, 973-4-5.   (633-4-5, 952-3-5)
    Coronary sinus:—
        Great coronary or cardiac vein :—
            Anterior interventricular vein.
            Left auricular vein.
            Left marginal vein.
        Posterior, middle cardiac, or posterior interventricular vein.
        Right auricular, right coronary, or small coronary vein.

Coronary Sinus :—
      Right marginal vein, or anterior cardiac vein, or vein of Galen.
      Smaller anterior cardiac veins.
   Oblique vein.
Cardiac nerves, 975–6. (953–5)
   Deep cardiac plexus.
   Superficial cardiac plexus. Cardiac ganglion of Wrisberg.
   Coronary plexuses.
Lymphatics of the heart, 697. (682)
Peculiarities of the Fœtal Heart, 977–8. (955–6–7)
   Position.
   Weight.
   Auricular portion. Right auricle.
   Ventricular walls.
   Foramen ovale. Valve of the foramen ovale.
   Eustachian valve.
   Ductus arteriosus.
   The fœtal circulation.
The position of the Chief Orifices one to the other and to the Chest Wall, 972.
Fig. 554. (950–1)
   Pulmonary orifice. Aortic orifice.
   Pulmonary semilunar valves.
   Aortic semilunar valves.
   Tricuspid valve.
   Mitral valve.
   Muscular walls.

Study the structure of the trachea, bronchi, and lungs, and the arrangement of the vessels and air tubes in the lungs. Trace the pulmonary artery and its divisions, and the pulmonary vein and its tributaries.

The trachea :—
   Structure, 951–2. (930–1)
      Cartilaginous rings. Fibrous membrane.
      First cartilage.
      Last cartilage.
      Fibres of the trachealis.
      Yellow elastic fibres.
      Mucous membrane.
      Arteries.
      Veins.
      Nerves.

The bronchi, 952. (931)
   Right bronchus.
   Left bronchus.   Bronchioles.
The lungs, 961–2. (940)
   Weight of the lungs.
   Color.
   Structure :—
      Serous coat.
      Subserous layer.
      Parenchyma :—
         Minute lobules :—
            Bronchial tube with terminal air cells.
            Pulmonary and bronchial vessels, nerves, and lymphatics.
               The vessels :—
                  Pulmonary arteries.
                  Radicles of the pulmonary veins. Arterial blood.
                  Bronchial arteries.
                  Bronchial veins :—
                     Superficial.

The Lungs—The Vessels:—
>>> Bronchial veins—Deep set.
>>> Lymphatics.
>>> Nerves.

The œsophagus :—
> Structure, 988. (967)
>>> Muscular, submucous, and mucous coats.
>> Muscular coat :—
>>> Longitudinal fibres.
>>> Circular fibres.
>> Submucous coat.
>> Mucous coat.

The Articulations at the Front of the Thorax, 222-3-4-5.

> The Chondro-sternal Articulations.
>>> Class :—*Diarthrosis.* Subdivision :—*Ginglymus.*
>> Anterior chondro-sternal ligament.
>> Posterior chondro-sternal ligament.
>> Superior and inferior ligaments.
>> Interarticular ligament.
>> Chondro-xiphoid ligament.
>> Synovial membranes.
>> Arterial supply.
>> Nerves.
>> Movements.
> The Costo-chondral Joints.
>>> Class :—*Synarthrosis.*
> The Union of the Segments of the Sternum with one another.
> The Interchondral Articulations.
>>> Class :—*Diarthrosis.* Subdivision :—*Arthrodia.*
>> Capsule.
>> Arteries. Nerves.
>> Movements.

The Articulations of the Ribs with the Vertebræ, 218-19-20-1-2. (225-7-8-9)

> The Costo-central Articulation.
>>> Class :—*Diarthrosis.* Subdivision :—*Ginglymo-Arthrodia.*
>> Capsular ligament.
>> Interarticular ligament.
>> Anterior costo-central or stellate ligament.
>> Synovial membranes.
>> Arterial supply.
>> Nerve-supply.
>> Movements.
> The Costo-transverse Articulation.
>>> Class :—*Diarthrosis.* Subdivision :—*Arthrodia.*
>> Capsular ligament. Eleventh and twelfth ribs.
>> Middle costo-transverse or interosseous ligament.
>>> Eleventh and twelfth ribs.
>> Posterior costo-transverse ligament.
>>> Eleventh and twelfth ribs.
>> Superior costo-transverse ligament ; from crest.
>>> First rib.
>> Synovial membrane.
>> Arterial and nerve supplies.
>> Movements.

MOVEMENTS OF THE THORAX AS A WHOLE, 225-6. (232-3)

The Articulations of the Vertebral Column, 201 to 208. (209 to 216)
  (*a*) Those between the bodies and intervertebral discs which form amphiarthro-
    dial joints.
  (*b*) Those between the articular processes which form arthrodial joints.
  Immediate :—
    (*a*) Those between the bodies and discs.
    (*b*) Those between the articular processes.
  Intermediate :—
    (*c*) Those between the laminæ.
    (*d*) Those between the spinous processes.
    (*e*) Those between the transverse processes.
  The Articulations of the Bodies of the Vertebræ.
        Class :—*Amphiarthrosis.*
    Intervertebral substances :—
      Laminar portion.
      Central portion.
    Anterior common ligament.
    Posterior common ligament.
    Lateral or short vertebral ligaments.
  The Ligaments connecting the Articular Processes.
        Class :—*Diarthrosis.* Subdivision :—*Arthrodia.*
    Capsular ligaments.
    Synovial membrane.
  The ligaments uniting the Laminæ.
    Ligamenta subflava.
  The Ligaments connecting the Spinous Processes with one another.
  The ligaments connecting the Transverse Processes with one another.
    Supraspinous ligament.
    Interspinous ligaments.
    Intertransverse ligaments.
  Arterial supply.
  Nerve-supply.
  Movements.
      Neck.
      Thoracic region.
      Lumbar region.

The Male Perinæum.

To expose the perinæum for dissection, draw the subject to the end of the table, flex the thigh and leg, and bind the palm of the hand over the dorsal surface of the foot. A slender stick placed between the knees will keep the thighs separated. The pelvis should be raised on a block.

THE PERINÆUM, 1150–1156. (1126–1132)

Bony boundaries.
    Two triangles :—(1) Anterior or urethral.
                    (2) Posterior or rectal.
  Central point of the perinæum, lateral lithotomy.
  The urethra.
  Ischio-rectal fossa.
    Anterior recess, posterior recess.
  Anus. Rectum.
  Dissection of the perinæum.

The rectum should be moderately distended with tow and the margins of the anus stitched together. A full-sized staff should be passed into the bladder and tied in place. The scrotum should be stitched forward out of the way.

The perinæum. The outlet of the pelvis, 1096. (1072–3)
The male perinæum, 1096. (1073–4)
    The ischio-rectal region, 1097. (1074)

Beginning at the base of the scrotum, make an incision along the median raphe, around the margin of the anus and back to the end of the sacrum. Make a transverse incision passing just in front of the tuberosities of the ischium. Carefully reflect the skin and expose the superficial fascia.

In the urethral triangle the superficial fascia consists of two well-marked layers, a superficial layer and a deeper layer, or fascia of Colles. The deep layer, or fascia of Colles, turns around the posterior border of the superficial transverse perinæi muscles, and blends with the posterior border of the triangular ligament, not extending over the rectal triangle. The superficial layer of the fascia should now be removed, carefully preserving the deeper layer, or fascia of Colles. In the rectal triangle the external sphincter ani and the superficial vessels and nerves will be exposed.

Sphincter ani externus, action, 1097-8. (1074)

Dissect out the fat from the ischio-rectal fossæ, and expose the hæmorrhoidal vessels and nerves. The greatest care should be exercised not to injure the fascial boundaries of the ischio-rectal fossæ.

Ischio-rectal fossæ, contents, 1102-3. (1097)
    The obturator fascia, 1098-9 ; Fig. 648. (1075)
        Lower or ischio-rectal segment.
    Ischio-rectal or anal fascia, 1102 ; Fig. 648. (1077-9)
External or inferior hæmorrhoidal branches, 609. (597)
Inferior hæmorrhoidal nerve, 858. (838)
Perinæal branch—of the fourth sacral nerve, 853. (834)
Superficial lymphatics of the perinæum, 700. (685)
Perinæum proper, fascia of Colles, 1103. (1080)
    Fascia of Colles, 1103-4. (1080-1)

Demonstrate the attachments of the fascia of Colles, and the extent and outline of the superficial perinæal interspace. To do this, make a short incision through the fascia, a little to one side of the median line ; introduce the point of the blowpipe, and drive air into the space ; the attachments of the fascia and the outline of the space will be rendered distinct.

Divide the fascia of Colles and reflect it, exposing the structures in the superficial interspace. Fig. 370 A, P. 609, and Fig. 371. (598)

Superficial perinæal interspace, 1104-5. (1081)
    Contents :—(1) The crura, 1070. (1047)
        The ischio-cavernosus, 1072. (1049)
            Compressor venæ dorsalis.
    (2) The bulb, 1071-2. (1049)

Expose the bulbo-cavernosus; study its structure and relations, then divide the muscle near the median line and reflect it from the surface of the bulb.

    Bulbo cavernosi, 1072. (1049)
    (3) Superficial transversi perinæi, 1072. (1049-50)
    (4) Arteries of the corpora cavernosa.
    Dorsal arteries of the penis.
    Veins. Lymphatics.
        Artery of the corpus cavernosum (crus), 611. (599)
        Dorsal artery of the penis, 611. (599)
    (5) Dorsal nerves of the penis, 858. (837)
    (6) Superficial perinæal vessels and nerves.
        Superficial perinæal branch, 609-10. (597-8)
        Transverse perinæal artery.
        Posterior or external superficial perinæal nerve, 858. (838)
        Anterior or internal superficial perinæal nerve, 858. (838)

The long pudendal nerve may be traced to advantage at this time. Follow the nerve back to its superficial origin and trace the branches forward.

The long pudendal nerve, 857. (837)

Remove the muscles from the superficial perinæal interspace and expose the inferior triangular ligament.

Inferior triangular ligament, 1105. (1081-2)

Divide and turn off the inferior triangular ligament, exposing the deep perinæal interspace. See Fig. 370, P. 609 (left side). (598)

Deep perinæal interspace, 1105-6. (1082-3)
Contents:—(1) Membranous urethra, 1074. (1052)
(2) Cowper's glands.
(3) Internal pudic arteries.
Artery of the bulb, 610-11. (598-9)
(4) Pudic veins.
(5) Pudic lymphatics.
(6) Dorsal nerve of the penis.
(7) Transversus perinæi profundus, or deep transversus perinæi or compressor urethræ.
Muscular division of pudic nerve, 858. (838)
Superior or deep triangular ligament, 1106. (1083)
Reflect the ischio-rectal fascia from the parietal surface of the levator ani muscle; the muscle will be considered again at a later period of the dissection.
Levator ani muscle, 1099-1100. (1075)

Female External Genitals and Perinæum.

To expose the perinæum for dissection, draw the subject to the end of the table, flex the thigh and leg, and bind the palm of the hand over the dorsal surface of the foot. A slender stick placed between the knees will keep the thighs separated. The pelvis should be raised on a block.

Female external genitals, 1156-7-8-9. · (1132-4-5); also, 1076-7. (1053-4)
Labia majora. Labia minora or nymphæ. Vestibule.
Vaginal orifice, hymen, carunculæ myrtiformes.
Fourchette and fossa navicularis.
Examination per vaginam.
The perinæum, 1150-1. (1126)
Bony boundaries.
Two triangles:—(1) Anterior or urethral.
(2) Posterior or rectal.
The rectum and vagina should be moderately distended with tow, and the orifice of each closed by stitching together the lateral margins.

The perinæum. The outlet of the pelvis, 1096. (1072-3)
Perinæum proper, fascia of Colles, 1103. (1080)
The female perinæum, 1107. (1083)
The ischio-rectal region, 1097. (1074)
Make elliptical incisions around the margins of the vagina and anus; make a median incision through the intervening space, from the pubis to the end of the sacrum. Make a transverse incision passing just in front of the tuberosities of the ischium. Carefully reflect the skin and expose the superficial fascia. In the urethral triangle the superficial fascia consists of two layers, a superficial layer and a deeper layer, or fascia of Colles. The deeper layer, or fascia of Colles, turns around the posterior border of the superficial transverse perinæi muscles, and blends with the posterior border of the triangular ligament, not extending over the rectal triangle. The superficial layer of the fascia should now be removed, carefully preserving the deeper layer, or fascia of Colles. In the rectal triangle the sphincter ani and the superficial vessels and nerves will be exposed.

Sphincter ani externus, action, 1097-8. (1074)
Dissect out the fat from the ischio-rectal fossæ, and expose the hæmorrhoidal vessels and nerves. Care should be exercised not to injure the fascial boundaries of the ischio-rectal fossæ.
Ischio-rectal fossæ, 1102-3. (1097)
Anterior recess, posterior recess.
Contents.
The obturator fascia, 1098-9; Fig. 648. (1075)
Lower or ischio-rectal segment.
Ischio-rectal or anal fascia, 1102; Fig. 648. (1077-9)
External or inferior hæmorrhoidal branches, 609. (597)
Inferior hæmorrhoidal nerve, 858. (838)
Perinæal branch—of fourth sacral nerve, 853. (834)
Superficial lymphatics of the perinæum, 700. (685)
Fascia of Colles, 1103-4. (1080-1)
Divide the fascia of Colles and reflect it, exposing the structures in the superficial perinæal interspace, Figs. 654 and 656.

7

Superficial perinæal interspace, 1104-5. (1081)
 Superficial perinæal branch, 609-10; Fig. 370 A (right side). (597-8)
  Transverse perinæal artery.
 Posterior or external superficial perinæal nerve, 858. (838)
 Anterior or internal superficial perinæal nerve, 858. (838)
 Muscles, 1079; also, 1107; Figs. 632, 656. (1056; also, 1084)
  Ischio-cavernosi (male), 1072. (1049)
  Bulbo-cavernosi, constrictor vaginæ.
  Transversi perinæi, 1072. (1049)
 Perinæal body, 1107-8. (1084)
 Vessels, 1077-8. (1054-5)

Carefully raise and remove the bulbo-cavernosi and transversi perinæi, and expose the bulbuli vestibuli and glands of Bartholin.

 Bulbuli vestibuli, 1078-9; Fig. 632. (1055-6)
  Pars intermedialis. Vessels, 1077-8. (1054-5)
 Glands of Bartholin, 1077. (1054)
 Clitoris, glans clitoridis, 1078. (1055)
 Inferior triangular ligament, 1105. (1081-2)
Deep perinæal interspace, 1105. (1082)
 The urethra, 1079. (1056)
 Deep transversus perinæi, 1107, 1106. (1084, 1082-3)
 In the female:—Artery of the bulb, 611. (598-9-600)
  Artery of the crus.
  Dorsal artery of the clitoris.
  Dorsal nerve of the clitoris, 858. (838)
Superior or deep triangular ligament, 1106. (1083)

Reflect the ischio-rectal fascia from the parietal surface of the levator ani muscle; the muscle will be considered again at a later period.

Levator ani muscle, 1099-1100. (1075)

THE ABDOMINAL WALLS.

Before beginning the dissection, study carefully the surface markings or landmarks of the region. With colored crayons indicate on the surface of the abdomen the position of each structure mentioned.

Superficial Anatomy of the Abdomen, 1141-2-3-4. (1118-19-21)
Skin markings; bones and muscular landmarks:—
 Linea alba. Linea semilunaris.
 Poupart's ligament, external abdominal ring.
  Internal ring. Canal.
 Vessels in the abdominal wall.
 Lymphatics.
 Nerves.
 The diaphragm.

For convenience in stating the position of the enclosed viscera, the abdomen is divided into nine regions. Outline the regions of the abdomen, and note the contents of each region, as indicated by the table.

THE ABDOMINAL VISCERA REGIONALLY ARRANGED, 1249. (1225)

Right.	*Middle.*	*Left.*
Hypochondriac.	Epigastric.	Hypochondriac.
Lumbar.	Umbilical.	Lumbar.
Inguinal.	Hypogastric.	Inguinal.

Viscera behind the linea alba, 1145-1150. (1121-1126)
  (1) Above the umbilicus.
  (2) Below the umbilicus.
 The liver. Gall bladder.
 Stomach. The pancreas.
 Intestines:—(A) Small.
   (B) Large intestine.

Landmarks for lumbar colotomy.
The kidneys.   The spleen.
Aorta and iliac arteries.
Some of the branches of the aorta :—
    Cœliac axis.   Superior mesenteric artery.
    Renal arteries.   Inferior mesenteric artery.

Dissection of the Abdominal Walls. Inguinal Hernia. Umbilical Hernia.

Make a circular incision and raise a button of skin around the umbilicus. With a narrow sharp knife, open into the abdominal cavity through the umbilicus; through this opening inflate the abdominal cavity enough to render the abdominal walls moderately tense, close the opening with a ligature tied under the circular button. Make a median incision from the middle of the sternum to the symphysis pubis, and two transverse incisions: one from a point midway between the pubis and umbilicus, to the anterior superior spine of the ilium and along the crest, the other from the umbilicus in the direction of the axilla, to the lower border of the pectoralis major. Beginning at the umbilicus turn the skin upward from that point, and outward from the median line, and expose the superficial fascia underneath.

The Abdominal Parietes, 425. (421-2)

Carefully turn off the first layer of the superficial fascia, in the same manner as the skin, and expose the superficial vessels and nerves.

The superficial nerve supply of the abdomen is derived from branches of the lower intercostal, ilio-hypogastric and ilio-inguinal nerves, Fig. 460, P. 844. (824)

Superficial nerves :—
  Lower intercostal nerves :—
    Lateral.cutaneous nerves of the abdomen, 845.   (826)
      Anterior branches.   Posterior branches.
    Anterior cutaneous nerves of the abdomen, 846.   (826)
    Last thoracic nerve, 846.   (826)
  Hypogastric branch—of the ilio-hypogastric, 848.   (829)
  Ilio-inguinal nerve, terminal branches, 849.   (829)
The superficial arteries :—
  From the intercostal arteries :—
    Lateral cutaneous branches, 583 ; Fig. 362.   (572)
  From the femoral : 618.   (606)

      The portion of the vessels above Poupart's ligament should be exposed ; the origin from the femoral will be shown later.

    Superficial external pudic.
    Superficial epigastric.
    Superficial circumflex iliac.
  Cutaneous branches—from the deep circumflex iliac, 613.   (602)
  Cutaneous branches—from the superior epigastric, 550.   (540)
  Cutaneous branches—from the deep epigastric, 613.   (601)
The superficial veins accompany the superficial arteries.
Lymphatics of the abdomen and pelvis, 699–700 ; Fig. 400.   (684)
  Parietal lymphatic vessels :—
      Superficial.   Deep.
  Superficial parietal lymphatics, front, lateral.
  The inguinal glands, 705.   (691)
    Oblique or inguinal glands proper.

Turn off the deep layer of the superficial fascia and expose the structures underneath.

Read : The Muscles, 296–302.   (299–305)
    1. The name.   2. The shape.   3. The attachments ; origin, insertion.
    4. The structure.   5. Nerve-supply.   6. Action.
The abdominal muscles, linea alba, umbilicus, 425–6.   (422)
Obliquus externus abdominis, 428–9–30 ; Fig. 303.   (424–6–7)
  Origin.   Insertion.
  Structure :—Triangular fascia, Fig. 285 ; P. 365.   (364)
      External abdominal ring.
      Poupart's ligament.   Gimbernat's ligament.

External abdominal ring:—
>   Inner pillar.   External pillar.   Intercolumnar fibres.
>   Intercolumnar fascia.
Nerve-supply.   Action.   Relations.   Variations.

Carefully divide the aponeurosis of the external oblique by a transverse incision, from near the anterior superior spine of the ilium to the outer border of the rectus, then downward from that point to the pubis, passing to the inner side of the external abdominal ring in order to preserve it intact; turn the flap thus formed down and out to Poupart's ligament. Also divide the muscle at its origin and near the crest of the ilium, and turn forward to the point where the aponeurosis blends with that of the internal oblique. The ilio-hypogastric—848—(829) and ilio-inguinal—849—(829) nerves perforate the internal oblique near the crest of the ilium and cross the lower portion of the muscle.

Obliquus internus abdominis, 431–2.   (427–8).   Fig. 266; P. 312.   (315)
>   Origin.   Insertion, conjoint tendon.
>   Structure, linea semilunaris.
Nerve-supply.   Action.   Relations.   Variations.
Cremaster, 432.   (428).   Fig. 285; P. 365.   (364)
>   Origin.   Insertion.   Structure.   Nerve-supply.   Action.   Relations.
Genital branch—of the genito-crural nerve, 849.   (829)

Divide the internal oblique near its origin from the crest of the ilium and at Poupart's ligament, detach from the ribs above, and by a vertical incision, from the lumbar fascia behind; care must be exercised to divide the internal oblique only. Near the crest of the ilium the branches of the deep circumflex iliac artery lie between the internal oblique and transversalis, and will indicate the depth of the incision required to divide the internal oblique. Passing forward upon the transversalis will be seen portions of the lower intercostal, ilio-hypogastric, and ilio-inguinal nerves.

Nerves:—
>   Lower intercostal nerves:—Branches, 845.   (826)
>   Last thoracic nerve, 846.   (826)
>   Ilio-hypogastric nerve, 848.   (829)
>   Ilio-inguinal nerve, 849.   (829)
Arteries:—
>   From the lower intercostal:—
>>   The intercostal portion, 582.   (571)
>>   Subcostal or twelfth dorsal, 583.   (572–3)
>   Lumbar arteries, 587–8.   (576–7)
>   Deep circumflex iliac, muscular branches, 613.   (601–2)
Veins accompany the arteries.
Lymphatics:—The deep parietal lymphatics—of the anterior and lateral abdominal
>   walls, 700.   (685)
Transversalis abdominis, 433–4–5; Fig. 304.   (429-30)
>   Origin.   Insertion.
>   Structure; anterior aponeurosis, posterior aponeurosis or lumbar fascia.
Nerve-supply.   Action.   Relations.
Sheath of the rectus, front portion of sheath, 435.   (430–1)

Open the sheath of the rectus by a vertical incision in the median line of the sheath, reflect the outer portion of the sheath to the outer border of the muscle and the inner portion toward the linea alba. The pyramidalis and rectus will be exposed.

Pryamidalis, 426.   (422–3) Fig. 266; P. 312.   (315)
>   Origin.   Insertion.   Structure.   Nerve-supply.   Action.   Relations.   Variations.
Rectus abdominis, 426–7–8.   (423–4)   Fig. 266; P. 312.   (315)
>   Origin, outer head, inner head.   Insertion.   Structure, lineæ transversæ.
Nerve-supply.   Action.   Lineæ transversæ.   Relations.   Variations.

Divide the rectus transversely near the umbilicus, turn the lower portion down to the origin and the upper portion to the insertion, expose the structures in or under the muscles. In raising the muscle note the anterior cutaneous nerves entering its deep surface.

Deep epigastric artery, 612–13.   (601)
>   Branches:—Cremasteric.   Pubic.   Muscular.   (Cutaneous.)   Terminal.
Superior epigastric artery, 550; Fig. 344.   (540)
>   Branches:—Muscular.   Xiphoid (peritoneal).

Posterior portion of the sheath, 435. (431)
    Fold of Douglas.   Transversalis fascia.

Divide the transversalis at its origin from Poupart's ligament and carefully raise and turn forward a small portion of the lower part of the muscle, exposing the transversalis fascia.

Transversalis fascia, 436.   (431–2)
    Deep crural arch.   Internal abdominal ring.   Infundibuliform fascia.

At the lower part of the abdomen, the parts should now be carefully replaced and their relations considered with reference to the occurrence of inguinal hernia.

Parts Concerned in Inguinal Hernia, 1159 to 1165.   (1135 to 1141)
External Ring :
    Formation.   Boundaries.   Shape.
    Intercolumnar fascia.   External spermatic fascia.
    Effect of the Position of the Thigh on the Ring.
Inguinal Canal :
    Length.   Direction.   Boundaries.
Internal Ring :
    Site.   Shape.   Dimensions.   Boundaries.
Forms of inguinal hernia :
    (A) The Common Form : External, or Oblique.
    (B) Rarer Form : Internal, or Direct.
A. Oblique, External Inguinal Hernia, coverings of :
            (1) At the internal ring, or inlet.
            (2) In the canal.
            (3) At the external ring, or outlet.
B. Direct or Internal Inguinal Hernia, coverings of :

Make a median incision through the abdominal wall, from the lower end of the sternum to the umbilicus.   Make a transverse incision just above the umbilicus.   Draw the flaps thus formed out and up, and stitch them in place.   Raise the lower portion of the abdominal wall and examine the inner surface in the inguinal region.

Posterior aspect—of inguinal hernia :
            Cords and fossæ.
            Three cords :—Urachus.
                        Obliterated hypogastric arteries.
            Three fossæ :—Internal, middle fossa, external fossa.
Causes of Hernia :
    (1) Hereditary.
    (2) Weak Spots.
    (3) (4) (5) (6) (7) (8) (9) (10) (11)

Parts Concerned in Umbilical Hernia, 1169-70.   (1145-6)
    Congenital umbilical hernia.
    Infantile umbilical hernia.
        Prevented by :—Changes in the Ring Itself.
                    Changes in the Vessels Themselves.
    Umbilical hernia of adult life.
    Coverings of an umbilical hernia.

Continue the division of the abdominal wall to the pubis, carrying the incision one-half an inch to the left of the median line.   Note again the position of the obliterated hypogastric arteries and the urachus.   The scrotum and spermatic cord will next be considered.

Make an incision from the external ring to the tip of the scrotum, in line with the spermatic cord, reflect the skin and expose the dartos.   Demonstrate the septum scroti.

The Scrotum, 1060-1.   (1038–9)
    Scrotal integument, raphe.
    Superficial lymphatics of the scrotum, 700.   (685)
    Dartos.
        Septum scroti.

On the right side divide the spermatic cord at the external ring, carefully remove the cord and testicle, with their coverings, and pin them out in a dissecting tray. The coverings of the cord and testicle can be readily demonstrated by dissection under water.

On the left side, reflect the coverings, and expose the cord in place; demonstrate and trace its constituent parts. Follow the vas deferens from the testicle to the brim of the pelvis. Drop the testicle in the pelvic cavity, not dividing the vas deferens, and preserve it for study later.

External spermatic or intercolumnar fascia.
Cremasteric or middle spermatic fascia.
Internal spermatic or infundibuliform fascia.
Tunica vaginalis.
Vessels and nerves:
    Arteries. Veins. Lymphatics.
    Nerves.

Spermatic Cord, 1068-9. (1045-6)
Vas deferens.
Spermatic artery:
    Branches:—Epididymal and testicular, 596. (584)
Spermatic veins:
    Pampiniform plexus, 671. (656-7)
Lymphatics.
Sympathetic nerve.
Processus vaginalis.
Internal cremaster.
Fat and connective tissue.

Cut away the superficial muscles of the abdominal wall, stitch back the flaps formed by the division of the walls, and examine the abdominal organs.

The Abdomen and Its Contents.
The Abdomen, 994-5. (973-4)

Study the location of the organs in the abdominal cavity as indicated by the table:—

The Abdominal Viscera Regionally Arranged, 1249. (1225)

The general description of each organ and its relations to the surrounding parts, should now be carefully considered, while the organs are in place. The structure, etc., will be studied later.

The stomach, 996-7-8; Figs. 566-7. (974-5-6-7)
    Cardiac orifice.
    Pyloric orifice, or pylorus.
    Borders. Surfaces.
    Relations to surrounding parts.
    Relations to the peritoneum.
    Alteration of position.
The small intestine, 1000. (978)
        (Duodenum.) Jejunum. Ileum.
    Jejunum and ileum, 1002-3. (981-2)
        Jejunum.
        Ileum. Meckel's diverticulum.
    The mesentery.
The large intestine, 1005-6-7. (983-4-5-6)
    Cæcum, or caput coli.
        The vermiform appendix.
        Ileo-cæcal fossa.
    The colon, 1008-9. (986-7)
        The ascending colon.
        The transverse colon.
        The descending colon.
    The sigmoid flexure and rectum, 1009-10-11. (987-8-9)
The liver, 1012-13-14-15. (990-1-2-3)
    Weight. Borders. Extremities. Surfaces. Lobes. Fissures. Ligaments.
    Anterior border. (Posterior border.)

The Liver:—
    Right extremity.   Left extremity.
    Superior surface.
    Inferior surface.
        Longitudinal fissure.
        Umbilical fissure.
        Quadrate lobe.
        Gall bladder, 1020–1. (998)
            Fundus. Neck. Body.

    General position, 1016–17-18. (994-5-6)
    Ligaments of the liver, 1019. (997)
        Coronary ligament.
        Right and left lateral ligament.
        Broad ligament. Round ligament.
The spleen, 1025–6–7–8. (1003–4–5)
    Position. Shape.
    Surfaces :—
        External or posterior surface.
        Anterior.
        Inner or renal surface.
    Anterior border. Posterior border.
    Size. Varieties.
    General relations of the spleen. (1006)

The general location of the kidney may now be determined, its exact position and relations will be fully considered later.   The pancreas will also be exposed at a later period of the dissection.

The kidneys, 1042–3–4–5. (1020–1–2–3)
    (Position and relations.)

Carefully trace the peritoneum and demonstrate its attachments, reflections, and relations to the abdominal walls and abdominal organs.   Study the plan of formation of the omenta, mesenteries, and ligaments.

The peritoneum, 989–90–1–2–3–4. (968–9–70–1–2)
    Greater and lesser sacs of the peritoneum.
    Foramen of Winslow.
    Course of the peritoneum in a longitudinal section of the body.
        Recto-vesical pouch. Recto-vaginal pouch.
        Lesser sac. Great omentum. Lesser or gastro-hepatic omentum.
        Gastro-splenic omentum.
        The gastro-phrenic and phreno-colic ligaments.
    Fossa duodeno-jejunalis, 1002. (981)

In connection with the demonstration of the peritoneum, the student should also read the chapter on The Evolution of the Peritoneum and an Explanation of its arrangement in the Human Body. Pages 1028–1041.   (1006–1019)

The structures between the layers of peritoneum forming the mesentery and transverse meso-colon should now be exposed. To do this, raise the great omentum and transverse colon and throw them up over the lower wall of the chest. Draw down the mass of small intestines and spread them out in such a manner as to expose the anterior peritoneal layer of the mesentery. Beginning at the upper part of the jejunum and proceeding downward, strip off the anterior layer of peritoneum from the mesentery of the small intestine, also the peritoneum covering the ascending colon, and the inferior layer of the transverse mesocolon. Expose the mesenteric artery and trace its branches. The mesenteric veins, nerves, the mesenteric glands and lacteals, will also be exposed at the same time.

Superior mesenteric artery, 592–3–4; Fig. 366. (581–2–3)
    Branches:—Intestinal branches, or vasa intestini tenuis.
            Ileo-colic.
            Right colic.
            Middle colic.
            Inferior pancreatico-duodenal.
    Variations in the superior mesenteric artery.

Superior mesenteric plexus, 871. (852)

Superior mesenteric vein, 674–5 ; Fig. 393.   (661)
    Tributaries :—Right gastro-epiploic.
               Pancreatico-duodenal veins.
Lymphatic vessels of the intestines, 703.   (687)
    Lymphatics of the small intestine, lacteals.
    Lympathics of the large intestine, (a), (b).
Lymphatic glands of the intestines, 703.   (687–8)
    Mesenteric glands.
    Meso-colic glands.

Expose the inferior mesenteric vessels; to do this, draw the small intestine to the right, carefully strip off the peritoneum from the lower portion of the aorta, and to the left, beyond the descending colon and sigmoid flexure.

Inferior mesenteric artery, 596–7 ; Fig. 367.   (586–7)
    Branches of the inferior mesenteric :—
               Left colic.
               Sigmoid artery.
               (Superior hæmorrhoidal.)

    Chief variations in the inferior mesenteric.

Inferior mesenteric plexus, 871.   (852)
Aortic plexus, 871.   (852)
Inferior mesenteric vein, 676 ; Fig. 394.   (662)
Lymphatics of the large intestine, (c), 703.   (687)

Remove the jejunum and ileum from the abdominal cavity.   First tie two ligatures around the jejunum at its commencement, and divide the intestine between them; divide the ileum in the same manner, about four inches above its termination in the cæcum.   Divide the mesentery near the intestine.   The portion of intestine thus removed should be washed out and carefully dissected to show the structure.

Structure of the small intestine, 1003–4–5 ; Figs. 573–4.   (982–3)
    Blood-supply of the small intestine.
    Lymphatics and nerves.

Ligature the sigmoid flexure at the brim of the pelvis, and divide the intestine at that point; divide the great omentum along the lower border of the stomach below the gastro-epiploic arteries.   Remove the remainder of the ileum and the colon, wash them out, and study the structure.   The lower portion of the ileum and the first six inches of the large intestine should be inflated and allowed to dry; on the dried specimen, remove the lateral wall of the cæcum at the point of entrance of the ileum · the arrangement and structure of the ileo-cæcal valve will be shown.

Structure of the large intestine, 1012.   (990)
    Blood-vessels.   Nerves and lymphatics.
    The ileo-cæcal valve, 1007–8.   (986)

Expose the cœliac axis and its branches; to do this, raise the liver, draw down the stomach, and strip off the anterior layer of the gastro-hepatic omentum.   The gastric artery is easily found as it passes along the lesser curvature of the stomach, and by tracing it to its origin the cœliac axis can be readily exposed and its divisions followed.

Cœliac artery or cœliac axis, 588–9 ; Fig. 365.   (577–8)
    Relations :—In front, behind, above, below, right, left.
    Variations.

Branches of the cœliac artery, 589–90–1–2.   (578–9–80–1)
    Gastric or coronary artery :—
            Œsophageal branches.
            Cardiac branches.
            Gastric branches.
                Lesser anterior and posterior gastric branches.
                Greater anterior gastric.
            Hepatic branch.
        Chief variations.

Gastric or coronary vein, 674.   (660–1)
Hepatic artery.

Branches of the hepatic artery :—
  Pancreatic or lesser pancreatic.
  Superior pyloric.
  Gastro-duodenal :—
    Right gastro-epiploic :—
      Ascending or gastric branches,
      . Epiploic or omental branches.
    Superior pancreatico-duodenal.
    Inferior pyloric.
  Right terminal branch :—
    Cystic artery.
  Left terminal branch.

Chief variations.

Draw the stomach forward and throw the lower border upward on the chest wall ; retain it in this position while the splenic artery and its branches are exposed.

Splenic artery.
  Branches of the splenic artery :—
    Smaller pancreatic branches.
    Larger pancreatic branch.
    Left gastro-epiploic :—
      Ascending or gastric branches.
      Descending epiploic or omental branches.
    Vasa brevia.
    Terminal branches.

Variations in the splenic artery.

Cœliac plexus, 870–1.   (850)
  Splenic plexus.
  Hepatic plexus :—
    Pyloric.   Right gastro-epiploic.
    Pancreatico-duodenal plexus.
    Cystic plexus.
  Coronary plexus.
The Cœliac glands, 701.   (686)
Lymphatics of the stomach, 702.   (687)
  Superior gastric.
  Inferior gastric lymphatics.
  Left gastric lymphatics.
Lymphatic glands of the stomach, 702–3.   (687)
  Superior gastric glands.
  Inferior gastric or gastro-epiploic glands.
The splenic vein, 675–6.   (662)
  Tributaries :—Vasa brevia.
              Left gastro-epiploic.
              Pancreatic.
              Inferior mesenteric.
Hepatic glands, 704.   (688–9)
Portal vein, 673–4 ; Fig. 392.   (659–60)
  Sinus of the portal vein.
  Right branch.
  Left branch.
  Tributaries :—
    Pyloric.   (Gastric.)   Cystic.   (Superior mesenteric.   Splenic.)

Draw down the œsophagus and apply a ligature just below the diaphragm.   Loosen the ligature at the lower end of the duodenum, introduce the blowpipe and partially inflate the duodenum and stomach, then withdraw the blowpipe and renew the ligature.   Expose the duodenum and pancreas and study their relations.

Duodenum, 1000-1-2 ; Figs. 570-1.   (978-9-80-1)

> *First part, the superior or ascending.*
> *Second part, the descending portion.*
> *Third part, or transverse portion.*
> *Fourth part.   Musculus suspensorius duodeni.*

Pancreas, 1023-4-5 ; Fig. 587.   (1001-2)
>   Head of the pancreas.
>   Body of the pancreas.
>   Tail of the pancreas.

>   Open the gland by an incision a little below the median line, extending from the head to the tail, and expose the duct.

>   Duct of the pancreas.
>   Blood-supply.   Lymphatics.   Nerves.

The common bile duct, ampulla of Vater, 1021.   (999)
Abdominal branches of the pneumogastric nerve, 819.   (800)

Apply a ligature and divide the œsophagus just below the diaphragm.   Remove the stomach, duodenum, pancreas, and spleen, dividing the vessels and folds of peritoneum which hold them in place. The arteries should be divided about an inch from their origin.   Divide the pancreatic duct near the duodenum, remove the gland, and examine its structure.   Clean the surface, and wash out the duodenum and stomach, and study their structure.

The pancreas, 1023-4-5.   (1001-2)
>   Lymphatics of the pancreas, 704.   (689)

The stomach, 998-9 ; Figs. 568-9.   (977-8)
>   Structure.
>   Nerves.   ·Blood-supply.   Lymphatics.

The pylorus, 996 ; Fig. 566.   (974-5)
Duodenum-Structure of small intestine, 1003-4.   (982-3)
The spleen.
>   Structure.   Blood-supply.   Lymphatics.   Nerves, 1028.   (1005-6)
>   Lymphatics of the spleen, 704.   (689)
>   Splenic glands.

Remove the liver and demonstrate its parts and structure.   To remove the liver, cut the hepatic artery near its origin; divide the suspensory ligament from before backward, drawing the liver downward and forward; divide the lateral ligaments and the upper layer of the coronary ligament, separate the posterior border of the liver from the diaphragm, and divide the inferior layer of the coronary ligament. Divide the vena cava just above the posterior border, and again, just below it.   Cut the ligaments close to the abdominal wall, so that their connection with the liver can be studied when the organ is removed.

The liver.
>   Relation to the peritoneum, 1018-19.   (996-7)
>   Ligaments of the liver, 1019.   (997)
>   Coronary ligaments.
>   Right and left lateral ligaments.
>   Broad ligament.   .
>   Round ligament.

Carefully expose the fissures and demonstrate the vessels, ducts, etc., entering or emerging from the liver.

>   Inferior surface, posterior surface, fissures, 1014-15-16.   (992-3-4)
>   Longitudinal fissure :—
>   Umbilical fissure.
>   Fissure of the ductus venosus.
>   Transverse or portal fissure.
>   Fossa of the gall bladder.
>   Fissure of the vena cava.
>   Spigelian lobe.   Caudate lobe.
>   Gall bladder, 1020-1-2-3.   (998-9-1000-1)
>   Structure of the gall bladder.

Blood-vessels of the liver. Lymphatics, 1019-20. (997-8)
Hepatic veins, 672. (658)
Lymphatics of the liver, 703-4. (688)
    Superficial set of lymphatics.
        Upper or convex surface.
        Superficial lymphatics on the under surface of the liver.
    Deep set of lymphatics.
Structure of the liver, 1020. (998)
Varieties of the liver, 1023. (1001)

Trace the portal vein, hepatic arteries, and hepatic ducts in a portion of the liver; in another portion, trace the hepatic veins.

The kidneys:—
    Investment and fixation, 1043. (1021-2-3)

Remove the capsule of fat and connective tissue and expose the kidney and suprarenal body in place. Trace the vessels leading to and from them; and the ureter, from the kidney to the brim of the pelvis.

    Physical characters, 1042-3. (1020-1)
        Anterior or visceral surface.
        Posterior or parietal surface.
        Upper extremity.
        External border.   Internal border.
        Hilum.   Sinus.
        Varieties, 1050. (1028)
    Position and relations, 1043-4-5-6-7. (1023-4-5)
        Posterior surface.
        Upper extremity.
        Anterior visceral surface:—
            Right kidney.
            Left kidney:—
                Upper or gastric area.
                Middle or pancreatic area.
                Inferior or colic area.
        Outer border.   Inner border.
        Structures lying within the sinus.
Excretory duct of the kidney, 1051-2. (1029-30)
    Superior and inferior pelves.
    Common pelvis, ureter proper.
    Ureter:—
        Course and relations:—
            Abdominal portion.
        Lymphatics of the ureters, 705. (689)
Renal arteries, 594. (583)
    Branches:—Inferior suprarenal.
        Capsular or perineal branches.
        Uretal branch.

Variations in the renal arteries.

Renal plexus, 871. (850-2)
Renal or emulgent veins, 670-1. (656)

Suprarenal bodies, 1050. (1028)
    Right suprarenal body, anterior surface.
    Left suprarenal body, anterior surface:—
                Above.   Externally.
    Suprarenal lymphatics, 705. (689)
Suprarenal arteries, 595. (584)
    Superior suprarenals.
    Middle suprarenals.
    Inferior suprarenals.

Suprarenal plexus, 871. (850)
Suprarenal veins, 671. (656)
　　Right side.　Left side.

On the right side divide the renal and suprarenal arteries and veins, divide the right ureter at the brim of the pelvis and remove the right kidney and suprarenal body and study their structure. Leave the left kidney in place. To expose the structure of the kidney, divide it into two portions, Fig. 613, and divide one of these portions into two lateral halves, Fig. 609; from these sections some idea of the structure can be obtained. Divide the suprarenal body into lateral halves and examine its structure.

Structure—of the kidney, 1047–8–9–50. (1025–6–7–8)
　　Medulla.　Cortex.
　　Medulla :—
　　　Pyramids of Malpighi :—
　　　　Pyramids.　Papilla.　Calyx.　Foramina papillaria.
　　Cortex :—
　　　Cortex proper.
　　　Columnæ Bertini.
　　　Pyramids of Ferrein.
　　Uriniferous tubules, collecting tube.
　　Vessels :—
　　　Cortico—medullary arches :
　　　　Cortical arches.
　　　　Medullary.
　　　Efferent vessels.
　　　Lymphatics.　(Also 704.)　(689)
　　　Nerves.
Suprarenal bodies, 1050-1. (1028-9)
　　Structure :—
　　　　Cortex.
　　　　Medulla.
　　Vessels and nerves :—
　　　　Arteries.　Veins.　Lymphatics.
　　　　Nerves.

Expose the abdominal aorta, the common and external iliac arteries, also the inferior vena cava and iliac veins. The arterial and venous tributaries not already shown should now be exposed in the order indicated. In exposing the aorta, preserve the solar plexus and semilunar ganglia.

The abdominal aorta, 584-5. (573-4)
　　Relations :—In front.　Behind.　Right side.　Left side.

　　Variations.

　　Branches :—
　　　Right and left phrenic, 586-7. (575-6)
　　　　Right phrenic.
　　　　　Branches :—Anterior.　Posterior.
　　　　　　　Right superior suprarenal.
　　　　　　　Caval.　Hepatic.　Pericardial.
　　　　Left phrenic.
　　　　　Branches :—Anterior.　Posterior.
　　　　　　　Œsophageal.　Left superior suprarenal.
　　　　　　　Splenic.　Pericardial.
The diaphragmatic plexuses, 871. (850)
Phrenic veins or inferior phrenic veins, 672. (658)
Diaphragmatic lymphatics, 696. (681)
Spermatic arteries, 595-6. (584)
　　Branches :—Uretal.　Cremasteric.

　　Chief variations in the spermatic arteries.

Spermatic plexus, 871. (852)
Spermatic veins, 671. (656-7)
Lymphatics of the testicle, 702. (687)

Strip the peritoneum from the under surface of the diaphragm and expose that muscle.

Diaphragm, 423-4-5. (419-20-1)
    Origin :—Anterior or sternal portion.
        Lateral or costal portion.
        Posterior or vertebral portion :—
            Ligamentum arcuatum externum.
            Ligamentum arcuatum internum.
            Crus ; right side, left side.
    Insertion.
    Structure :—*Central tendon.*
        Foramina.
    Nerve-supply. Action. Relations. Variations.
Iliac fascia, 368-9. (368) Lumbar fascia, 434. (430)

Carefully reflect the fascia from the surface of the psoas, iliacus, and quadratus lumborum. In reflecting the fascia, avoid injuring the branches of the lumbar plexus crossing these muscles.

Psoas magnus, 366-7. (365-6-7)
    Origin :—Inner part. Outer part.
    (Insertion.) Structure.' Nerve-supply. Action. Relations. Variations.
Psoas parvus, 368. (367-8)
    Origin. Insertion. Structure. Nerve-supply. Action. Relations.
Iliacus, 368. (367)
    Origin. (Insertion.) Structure. Nerve-supply. Action. Relations. Variations.

Carefully expose the branches of the lumbar plexus, tracing them from their origin to the point of exit from the abdominal cavity. To expose the nerves at their origin it will be necessary to dissect away a portion of the psoas muscle.

Lumbar nerves, 846. (826)
Lumbar plexus, 846-7. (827-8-9)
    Branches :—
        Ilio-hypogastric nerve, 848. (829)
        Ilio-inguinal nerve, 849. (829)
        Genito-crural nerve, 849. (829)
        External cutaneous nerve, 851. (832)
        Anterior crural nerve, 851-2. (832)
            Nerves to the iliacus.
        Obturator nerve, 849. (829-30)

        Accessory obturator nerve, 850. (831-2)

Quadratus lumborum, 435-6. (431)
    Origin. Insertion. Structure. Nerve-supply. Action. Relations.
Last thoracic nerve, 846. (826-7)
Subcostal artery or twelfth dorsal, 583. (572-3)
Subcostal vein or twelfth thoracic vein, 645. (633)
Lumbar arteries, 587-8. (576)
    Muscular branches. Renal branches.

    Variations.

Lumbar veins, 671-2. (657-8)
    Anterior branches.
    (Posterior branches.)

The Pelvic Viscera.

Review the relations of the peritoneum in the pelvic cavity and note the plan of formation of the peritoneal ligaments.

Recto-vesical pouch, 992. (971)
Bladder :—
    Peritoneum, false ligaments, 1056. (1034)

Female. {
Recto-vaginal pouch, 1056. (971)
Ligaments—of the uterus :—
    Peritoneal ligaments, 1085-6-7. (1062-3-4)
    Lateral or broad ligaments, or alæ vespertilionis.
    Superior or free border.
    Ligamentum infundibulo-pelvicum.
    Internal border.
    Inferior border.
    External border.
Structures enclosed between the two layers of the broad ligament :—
    (1) The ovary and its ligament.
    (2) Fallopian tube.
    (3) Round ligament.
    (4) Parovarium. Duct of Gärtner. Hydatid of Morgagni.
Posterior peritoneal or recto-uterine ligaments.
    Utero-sacral ligaments.
Anterior peritoneal or utero-vesical ligaments.
}

Strip the peritoneum from the pelvic walls, with the knife-handle carefully scrape away the subperitoneal tissue and fat and expose the upper or pelvic segment of the obturator fascia and the recto-vesical fascia. The ligaments of the bladder will be developed as the recto-vesical fascia is traced.

Subperitoneal connective tissue, 994. (973)
Obturator fascia, 1098-9 ; Fig. 684. (1075)
    White line or arcus tendineus.
    Upper or pelvic segment.
Recto-vesical fascia, 1101-2, 1056. (1077, 1034)
    Prostatic or anterior true ligaments.
    Lateral true ligaments.
    Superior true ligament.
The pelvic fascia, 1156. (1132)

The relations of the pelvic fascia should also be studied with a pelvis in which the fascia has been exposed from the side, by the removal of a part of the pelvic wall.

The position, relations, form, and size of each of the pelvic organs should now be determined, so far as is possible without dissection or injury to the structures.

The rectum. The anus, 1011. (989-90)
Urinary bladder, 1053-4-5-6. (1031-2-3)
    Form.
    Relations :—Antero-inferior surface. Cavum Retzi.
        Lateral surface.
        Posterior surface. Triangular space.
        Superior surface.
        In the infant, 1058. (1035-6)
Male. { Effects of distention. Internal urinary meatus. } 1055-6. (1033)
Female bladder, 1058. (1038)
Ureter, pelvic portion, 1052-3. (1030-1)
Male. {
Vas deferens, 1065-6. (1043-4)
Vesiculæ seminales, 1066-7. (1044-5)
Prostate, 1058-9. (1036-7)
    Base. Anterior wall. Posterior wall.
    Prostatic fissure.
}

Uterus or womb, 1081-2-3-4.   (1058-9-60-1)
   Isthmus.
   Body :—
      Anterior surface.
      Posterior surface.   Recto-vaginal pouch.
      Superior border.
      Lateral borders.   Broad ligaments.
      Superior angles.
      Fundus.
   Cervix :—Supravaginal zone.
          Zone of vaginal attachment.
          Intravaginal zone, or os uteri.   Labia.
   Dimensions.
   Direction. ·
   Variations in form according to age.

Female.

   Muscular ligaments, 1087-8.   (1064-5)
      Round or utero-inguinal.   Canal of Nuck.
      Utero-sacral ligaments.
      Utero-pelvic ligaments.
      Utero-ovarian.
  Ovary :—Form, position, etc., 1090.   (1066-7)
  Fallopian tubes, 1088-9.   (1065-6)
     Fimbriæ.   Fimbria ovarica.
     Isthmus.   Ampulla.
     Position.
  Vagina, 1079-80-1.   (1056-7-8)
    Form and direction.
    Relations :—
      Anteriorly.
      Posteriorly.   Pouch of Douglas.
      Laterally.   Duct of Gärtner.

The internal iliac artery and its branches should now be exposed, and the branches followed to the organ or part to which they are distributed, or to the place of exit from the pelvis.

Internal iliac artery, 600.   (589)
   Relations :—
     Behind.
     In front.

Variations.

Branches of the anterior division :—
    Hypogastric artery, 604.   (593)
    Vesical arteries, 604-5.   (593)
      Superior vesical :—
        Deferential or artery of the vas deferens.   (Male.)
        Uracheric branch.
        Ureteric branches.
      Middle vesical.
      Inferior vesical.
        Accessory pudic.

Female.
{
Ovarian artery, 596, 1091-2. (585, 1068)
   Branches :—Uretal.
         Fallopian.
         Uterine.
         Ligamentous.
Funicular artery, 1092. (1068)
Uterine artery, 605-6, 1091. (594, 1067-8)
   Branches of the uterine artery :—
      Cervical.
      Vaginal azygos.
Vaginal arteries, 606. (594-5)
   Azygos artery of the vagina.
Uterine veins. Ovarian veins, 1091-2. (1068)
}

Middle hæmorrhoidal or middle rectal artery, 605. (593-4)

Obturator artery, 606-7. (595)
   Iliac or nutrient branch.
   Vesical.
   Pubic branch.
Sciatic artery, 607. (595-6)
   Intrapelvic branches.
Internal pudic artery—within the pelvis, 608. (597)
Branches of the posterior division of the internal iliac artery :—
Ilio-lumbar artery, 601-2. (589-90-1)
   Iliac branch.
   Lumbar branch.
Gluteal artery, 602-3. (591)
   Branches—within the pelvis.
Lateral sacral arteries, 602. (591)
   Superior artery.
   Inferior lateral sacral :—
      Posterior or spinal branches.
      Anterior or rectal branches.
      External branches.
      Internal branches.
Middle sacral artery, 598.
   Lateral sacral branches.
   Rectal or hæmorrhoidal branches.

   Variations.

   Fifth pair of lumbar arteries, 688. (577)
   Middle sacral veins, 672-3. (659)
Veins of the pelvis, 676-7-8. (663-4)
   Internal iliac vein :—
      Tributaries :—Gluteal veins. Ilio-lumbar veins. Lateral sacral veins.
         Obturator vein. Sciatic veins. Pudic vein.
         Prostatico vesical plexus (male). Vesical plexus.
         Hæmorrhoidal plexus of veins :—
            Inferior, middle, superior.
Lymphatics of the pelvic viscera, 702. (686)
   Lymphatics of the bladder.
   Lymphatics of the rectum.
   The lymphatics of the uterus, vagina, ovaries, and Fallopian tubes.
Pelvic walls, 700. (685)

Draw the pelvic organs aside, strip off the recto-vesical fascia and expose the visceral surface of the levator ani and coccygeus muscles.

Levator ani muscle, 1099-1100. (1075-6)
Coccygeus, 1100-1. (1076-7)
8

Draw the pelvic organs forward and expose the sacral and coccygeal nerves. Trace the branches to their distribution or to the place of exit from the pelvic cavity.

Sacral and coccygeal nerves, 853; Fig. 465. (833)
　　　Fourth sacral nerve :—
　　　　　Perforating cutaneous branch.
　　　　　Perinæal branch. ·
　　　　　Branches to coccygeus and levator ani.
　　　Coccygeal plexus, 854. (834)
　　　　　Anterior branches.
　　　　　Posterior branches.
　　　　　Coccygeal nerve.
Sacral plexus; great sciatic nerve, pudic nerve, 854; Fig. 465. (834)
　　"Branches of the sacral plexus," collateral and terminal.
　　　　　Collateral branches :—
　　　　　　　Superior gluteal nerve, 854-5. (835)
　　　　　　　Inferior gluteal nerve, 855. (835-6)
　　　　　　　Nerve to the pyriformis, 855. (836)
　　　　　　　Visceral branches.
　　　　　　　Nerve to the quadratus, 855-6. (836)
　　　　　　　Small sciatic, 856-7. (837-8)
　·　　　　　　Nerve to the obturator internus, 857. (837-8)
　　　　　Terminal branches :—
　　　　　　　Great sciatic nerve, 858. (838-9)
　　　　　　　Pudic nerve, 857-8. (838)

Expose the hypogastric plexus and the sacral portion of the gangliated cord and note the branches of communication with the sacral plexus. So far as possible, trace the branches of distribution to the pelvic viscera.

Hypogastric plexus, 871. (852)
　　Pelvic plexuses, 873. (852-3)
　　　　Middle hemorrhoidal plexus.
　　　　Vesical plexus:—
　　　　　　Superior group.
　　　　　　Inferior group.
　　　　　　Nerves to the vas deferens (male).
　　　　Prostatic plexus (male).
　　　　　　Small cavernous nerve. °
　　　　　　Large cavernous nerve.
　　　　Vaginal plexus (female).
　　　　Uterine plexus (female).
Sacral part of the gangliated cord, 869. (849)
　　Rami communicantes.
　　Branches.
Male.—Suspensory ligament of the penis. Angle of the penis, 1070. (1047)

Remove the organs from the pelvic cavity and study their structure. To remove the organs, divide the vessels and nerves passing to them; draw the bladder and rectum away from the pelvic walls and carefully sever the connecting fascia and ligaments; (in the male; draw the penis downward and free it from its attachments to the pubic arch, cutting the crura and ligaments close to the urethra); finally, divide the levator and sphincter ani muscles. The left kidney and ureter should be removed with the bladder.

Carefully separate the rectum from the bladder (and prostate, in the male; uterus and vagina, in the female); inflate it and demonstrate the internal sphincter, and its coats.

The rectum, 1011-12. (989-90)
　　The anus.

Divide the left kidney by a vertical incision, extending from the outer border nearly to the hilum, opening the sinus and exposing the interior of the pelvis. Insert the blowpipe in the pelvis of the kidney and inflate the bladder through the ureter, a ligature being placed around the urethra. Clean the surface of the bladder (remove the bladder from the vagina and uterus, in the female), exposing the terminal portion of the ureters, also the vas deferens and the vesiculæ seminales, then carefully expose the prostate and the membranous portion of the urethra, with Cowper's glands. (In the female, expose the ureters and urethra.)

The ureter, pelvic portion, 1052-3. (1030-1)
Reproductive organs of the male, 1058. (1036)
The testicles, 1061-2-3-4-5. (1038-9-40-1)
    The testicle proper.
    The epididymis : —
        Globus major.
        Globus minor.
    Tunica vaginalis.
    Gubernaculum testis.   Internal cremaster of Henle.
    Processus vaginalis.
    Structure :—
        Tunica albuginea :—
            Mediastinum testis or corpus Highmorianum.
            Trabeculæ.
        Tunica vasculosa.
        Tubuli seminiferi, spermatozoa.
        Lobules.   Tubuli recti.   Rete testis.   Vasa efferentia.
        Epididymis :—
            Conus vasculosus.   Tube of the epididymis.   Vas deferens.
            Vas aberrans.
        Hydatid of Morgagni.
        Paradidymis, or organ of Giraldès.
Vessels and nerves of the testicle and its appendages, 1067-8.   (1045)
    Arteries.   Veins.   Lymphatics of the vas and testicle.
    Nerves.

Open the vas deferens by a short longitudinal incision about four inches above its termination, insert the end of a small blowpipe, and inflate the lower portion of the vas deferens and the seminal vesicle; they can then be readily exposed by careful dissection.

Vas deferens, 1065-6.   (1043-4)
    Ejaculatory duct.
Vesiculæ seminales, 1066-7.   (1044-5)
    Ejaculatory duct.
Urinary bladder :—
    Structure, 1056-7-8.   (1033-4-5)
        Muscular coat, diverticular sacculations.
        Submucous coat.

Open the bladder by a median vertical incision through the anterior wall, extending from the summit to the base; continue the incision along the median line of the urethra (dividing the prostate and opening the membranous portion of the canal, in the male). Spread out the bladder and display its inner surface.

        Mucous membrane.
            Trigone.   Plica ureterica.   Uvula of Lieutaud.
        Vessels :—
            Arteries.   Veins.   Lymphatics.
        Nerves.
Ureter, vesical portion, 1053.   (1031)

Insert a small probe in the lower portion of the ureter; with the point of the knife open the canal and trace it through the bladder wall.

    Structure :—
        Mucous membrane.
        Muscularis.
    Vessels and nerves :—
        Arteries.   Veins.   Lymphatics.
        Nerves.
    Varieties.
The female urethra, 1079.   (1056)

The urethra, 1073-4-5. (1050-1-2)
    Prostatic portion :—
        Collicus seminalis or verumontanum.
        Sinus pocularis or uterus masculinus.
        Ejaculatory ducts.
        Prostatic sinuses.
    Membranous portion.
Prostate :—
    Structure and function, 1060. (1037-8)
    Vessels and nerves :—
        Arteries. Veins. Lymphatics.
    Nerves.

Insert a small probe in the ejaculatory duct; with the point of the knife trace the duct to the vas deferens and the vesicula seminalis.

The penis, 1069-70-1-2. (1046-7-8-9)
    Corpora cavernosa.
    Corpus spongiosum.
    Root. Body. Neck. Glans.
    Coverings of the penis :—
        Prepuce. Frænum præputi.

Male. {

Remove the skin from the surface of the penis, expose and trace the dorsal vein, arteries, and nerves. Insert the point of a small blowpipe in the dorsal vein and distend with air.

Dorsal vein of the penis, 677. (663)
Dorsal artery of the penis, 611. (599)
Dorsal nerve of the penis, 858. (838)
Superficial lymphatics of the penis, 700. (685)
Dartos. Fascial sheath, 1070. (1047)

Reflect the fascial sheath, and expose the corpora cavernosa and corpus spongiosum.

Corpora cavernosa. Crus penis.
    Tunica albuginea. Septum pectiniforme.
Corpus spongiosum :—
    Glans :—Corona glandis. Neck. External urinary meatus.
    Body.
    Bulb.

Open the spongy portion of the urethra by a median incision, and examine the mucous membrane.

Urethra, spongy portion, 1075. (1052)
    Penile angle.
    External meatus.
    Lacunæ. Lacuna magnus.

Female Reproductive Organs: Structure.

Carefully clean and expose the uterus and vagina; spread them out, with the posterior surface uppermost, and demonstrate, first, the structure of the vaginal wall.

Vagina :—Structure, 1081. (1058)
    Fibrous coat.
    Muscular coat.

Open the vagina by a median incision through the posterior wall; make a short transverse incision at the upper end of the median incision, and reflect the vaginal wall outward, Fig. 635. Note the relations of the vaginal wall to the cervix.

Mucous membrane.
Vessels and nerves :—
    Arteries. Veins. Lymphatics.
    Nerves.

Uterus, 1083-4-5. (1060-1-2)
  Cervix :—
    Supra-vaginal zone.
    Zone of vaginal attachment.
    Intra-vaginal zone, os uteri.
      Labia.

Open the uterus by a Y-incision; beginning at the os, make a median incision through the posterior wall nearly to the fundus, and from the upper end of the median incision make incisions in the direction of the angles, to the termination of the oviducts, Fig. 637. Starting at the median incision, trim away the posterior wall of the uterus and expose the cavity.

Cavity :—
  Cavity of the body.
  Cavity of the neck.
    Os internum.
    Os externum.
Structure :—
  Muscular coat.
    Outer, middle layer, internal layer.
  Mucous membrane.
  Uterine artery. Uterine veins, 1091. (1067-8)
  Lymphatics, 1092. (1068-9)
Fallopian tube.
  Structure, 1089-90. (1066)
    Serous coat.
    Cellular coat.
    Muscular coat.
    Mucous membrane.
Ovary, 1090-1. (1067)
  Structure :—
    Tunica albuginea.
    Ovisacs or Graafian follicles. Corpus luteum.
  Ovarian artery. Ovarian veins, 1091-2. (1068)
Read :—Development of the Genito-urinary Organs, 1091-1096. (1069-1072)

The Thigh, 1210-1216. (1186-1192)
Bony landmarks :—
  Trochanter major. Nelaton's line. Bryant's triangle.
Muscular prominences. Poupart's ligament.
Scarpa's triangle. Saphenous opening.
Line of femoral artery. Hunter's canal.

The Knee, 1220-2-3-4. (1196-7-8-9-1200)
Bony landmarks :—
  The patella.
    In extension. In flexion. In semiflexion.
    Dislocation of the patella.
  The condyles and tuberosities.
  Ligamentum patellæ and tubercle of the tibia.
  Prepatellar bursa.
Synovial membrane ; Fig. 769.

Dissection of the Front of the Thigh.

With crayons indicate the distribution of the cutaneous nerves on the anterior aspect of the thigh, Fig. 464; P. 831.

Make an incision from the anterior superior spine of the ilium along the line of Poupart's ligament to the symphysis pubis and from the pubis backward, between the scrotum and the thigh, also, from the middle of Poupart's ligament down the middle of the thigh and knee to the lower end of the tubercle of the tibia. Make transverse incisions at the middle of the thigh, above and below the knee. Turn the skin each way from the median incision, exposing the superficial fascia, vessels, and nerves. Fig. 710; P. 1165.

Superficial fascia, 364. (363)
Superficial epigastric artery, 618. (606)
Superficial circumflex iliac artery, 618. (606)
Superficial external pudic artery, 618. (606)
Veins :—Each artery is accompanied by one or sometimes by two veins; these veins
open into the saphenous or femoral vein ; Fig. 397 ; P. 684. (670)
Superficial lymphatic glands, 705 ; Fig. 400. (691)
The inguinal glands :—
Oblique or inguinal glands proper.
Vertical set, or saphenous or superficial femoral glands.
Long or internal saphenous vein, 683-5 ; Fig. 397. (669)
Cutaneous nerves, P. 851 ; Fig. 464. (831)
External cutaneous nerve, 851. (832)
Posterior branch.   Anterior branch.
Crural branch—of the genito-crural nerve, 849. (829)
Terminal branches—of the ilio-inguinal nerve, 849. (829)
Middle cutaneous nerve, 852. (832)
Internal cutaneous nerve, 852. (832)
Anterior branch.   Posterior branch.
Branches of the long or internal saphenous nerve, 852-3.
Patellar branch.

Plexus patelle.   Subsartorial plexus, 852.   (833)

Turn off the superficial fascia and expose the deep fascia or fascia lata.

Deep fascia or fascia lata, 364-5 ; Fig. 285.
Ilio-tibial band.
Saphenous opening.   Iliac portion.   Falciform ligament.
Pubic portion.   Cribriform fascia.

Femoral Hernia, 1165-6-7-8-9.   (1141-2-3-4-5)
Parts concerned in femoral hernia :—
(1) Skin and superficial fascia of groin.
(a) Superficial layer of superficial fascia.
(b) Deep layer of superficial fascia.
(2) Poupart's ligament.
(3) Gimbernat's ligament.
(4) Fascia lata :—Iliac portion ; pubic portion.
(5) Saphenous opening.

Make an incision through the fascia lata from near the anterior superior spine of the ilium, along the
lower border of Poupart's ligament to the pubes, and from the same point, along the inner border of the
sartorius for about five inches.   Carefully reflect the fascia lata down and in and expose the femoral
sheath in place.

(6) Femoral sheath.

Beginning just below Poupart's ligament, make a vertical incision about two inches in length through
the femoral sheath, and expose the femoral artery ; in the same manner, expose the vein and the femoral
canal at the inner side of the vein.   Demonstrate the septum between the canal and the vein, and
between the vessels.

(7) Femoral canal :—
Length.   Limits.   Boundaries.   Contents.
(8) Femoral ring :—Shape.   Boundaries.
Position of vessels around the ring.
Course of femoral hernia.
Coverings of a femoral hernia :—
(A) At upper or femoral ring.
(B) In the canal.
(C) At the external or superficial opening.
Scarpa's triangle.

Carefully remove the deep fascia from the upper third of the thigh, exposing the muscles forming the
boundaries of Scarpa's triangle.   Dissect out the fat, fascia, and connective tissue, and expose the vessels
and nerves in the triangle, and the muscles forming its floor.

The femoral artery, 613–14–15–16; Fig. 372. (602–3–4, 606)
Relations of the common femoral artery :—
In front. Behind. Inner side. Outer side.
Relations of the superficial femoral artery in Scarpa's triangle :—
In front. Behind. Outer side. Inner side.
Branches of the femoral artery :—
    (A) From the common femoral, 618. (606)
        (1), (2), (3), already shown.
        (4) Deep or inferior external pudic artery.
        (5) Profunda or deep femoral (followed later).
    (B) From the superficial femoral in Scarpa's triangle, 621. (609)
        Muscular. Saphenous.

Reflect the fascia lata from the front and inner side of the thigh, knee, and upper portion of the leg. Expose and separate the muscles, as indicated. The vessels and nerves will be exposed as the muscles are separated, and should be traced to their place of distribution or followed through the region.

The sartorius, 369–70 ; Fig. 287. (368–9)
    Origin. Insertion. Structure. Nerve-supply. Action. Relations. Variations.

Follow the superficial femoral artery down the thigh, separating the muscles to expose the vessel. Note the relations of each portion.

Line of the femoral artery. Hunter's canal, 1213–14–15. (1189–90–1–2)

Open the canal and expose the vessels.

Relations of the superficial femoral artery in Hunter's canal, 616. (604)
    In front. Behind. Inner side. Outer side.
Branches of the superficial femoral in Hunter's canal, 621. (609)
        Muscular branches.
        Anastomotica magna :—
            Superficial branch.
            Deep branch.
The femoral vein, 686. (672)
    Tributaries. '
The common femoral vein, internally, 686. (672)
    Tributary.

Chief variations.

Deep inguinal or deep femoral glands, 707. (691)
Anterior crural nerve, 851–2–3 ; Fig. 463. (832–3)
    Branches:—Nerve to pectineus.
        Middle cutaneous nerve.
        Nerve to the sartorius.
        Internal cutaneous nerve.
            Anterior branch. Posterior branch.
        Long or internal saphenous nerve.
        Nerves to the quadriceps.
Tensor vaginæ femoris, 373 (372); Fig. 287, P. 370 (369); Fig. 303, P. 429. (425)
    Origin. Insertion. Structure. Nerve-supply. Action. Relations.
    Ilio-tibial band, 364 ; Fig. 287.
Quadriceps extensor femoris, 387 (385); Fig. 287, P. 370. (369)
    Rectus femoris, 387–8. (385–6)
        Origin. Insertion. Structure. Nerve-supply. Action. Relations.
    External circumflex artery, 619. (607)
        Ascending branch. Transverse branch. Descending branches.

Variations in the external circumflex.

Vastus externus, 388. (386)
    Origin. Insertion. Structure. Nerve-supply.
Vastus internus and crureus, 388–9–90. (386–7–8)
    Origin. Insertion. Structure.
    Ligamentum patellæ.

Vastus Internus and Crureus:—
    Nerve-supply.   Action.
    Relations of the vastus externus, crureus, and vastus internus.
    Variations.

The Inner Side of the Thigh.

The adductor muscles, 379; Figs. 287-9-90.   (377)
Gracilis, 383; (381-2)
    Origin.   Insertion.   Structure.   Nerve-supply.   Action.   Relations.
Adductor longus, 379; Fig. 289.   (378-9-80)
    Origin.   Insertion.   Structure.   Nerve-supply.   Action.   Relations.   Variations.

Divide the adductor longus just below its origin and turn the muscle downward and outward, exposing the profunda or deep femoral artery.

Profunda or deep femoral artery, 618-19.   (607)
    Relations:—Behind.   In front.   Externally.   Internally.
    Branches of the profunda:—(a), (b), (c).
            (c) Perforating arteries of the profunda, 620.   (608)
                Superior or first perforating; crucial anastomosis.
                Middle or second perforating.
                Inferior or third perforating.
                Fourth perforating.
Profunda or deep femoral vein, 686.   (672)
Pectineus, 370-1; Fig. 287.   (370)
    Origin.   Insertion.   Structure.   Nerve-supply.   Action.   Relations.   Variations.

Cut the pectineus at its origin and turn it downward and outward.

Anterior branch of obturator nerve, 849.   (830)

Accessory obturator nerve, 850.   (831-2)

Internal circumflex artery; crucial anastomosis, 619-20.   (607-8)

Variations of the internal circumflex.   .

Adductor brevis, 380-1-2; Fig. 289.   (380)
    Origin.   Insertion.   Structure.   Nerve-supply.   Action.   Relations.   Variations.

Cut the adductor brevis at its origin and turn downward and outward.

Posterior branch of obturator nerve, 849-50.   (831)
Adductor magnus, 382-3; Fig. 290.   (380-1)
    Origin.   Insertion.   Structure.   Nerve-supply.   Action.   Relations.   Variations.

Gluteal Region, 1216-17-19-20.   (1192-3-5-6)
Bony landmarks.
Gluteus maximus.
Nerves and vessels:—
    Superficial nerves, P. 1246; Fig. 791.   (1222)
    Great sciatic nerve.
    Gluteal artery.
    Sciatic and pudic arteries.

To expose the structures of the gluteal region: Make a median incision from the spine of the last lumbar vertebra to the tip of the coccyx, make a transverse incision from the same starting point to the
crest of the ilium and along the crest, also, a curved incision from the coccyx outward and downward
to the posterior median line of the thigh.   Remove the skin over the gluteal muscles and from the upper
third of the thigh and expose the superficial fascia and nerves.

Sacral and coccygeal nerves, 828.   (808-9)
    External branches (see internal branches).
    Lower two sacral, and coccygeal nerves.
Lumbar nerves, 827-8.   (808)
    External branches of upper three nerves.
Iliac branch of the ilio-hypogastric nerve, 848.   (829)
Iliac branch of last thoracic nerve, 846.   (826)   `

Posterior branch of external cutaneous nerve, 851. (832)
Reflected branches of the small sciatic nerve, 857. (837)
 The long pudendal nerve.
Perforating cutaneous branch of fourth sacral nerve, 853. (834)
Lymphatics of the gluteal region, 705. (689)

Remove the fascia, fat, and connective tissue and expose the gluteus maximus.

Gluteus maximus, 371-2-3 (370-1-2) ; P. 392 ; Fig. 291. (390)
 Origin. Insertion. Structure. Nerve-supply. Action. Relations. Variations.

Divide the gluteus maximus near its origin, beginning at the anterior border, and turn it downward and outward to the insertion ; much care is required to raise and turn off the muscle without injury to the underlying structures. A number of vessels and nerves will be seen entering the inner surface of the muscle ; these should be carefully divided close to the muscle, as it is reflected to the insertion.

The inferior gluteal nerve, 855. (835-6)
Superficial branch of the gluteal artery, 603. (591)
Gluteus medius, 373-4 ; Fig. 288. (372-3)
 Origin. Insertion. Structure. Nerve-supply. Action. Relations. Variations.

Beginning at the posterior border, divide the gluteus medius about two inches below the origin, turn the lower portion down to the insertion, carefully raise the upper portion to the origin and expose the structures underneath.

Deep branch of the gluteal artery, 604 ; Fig. 369. (592)
 Superior. Inferior branch.
Superior gluteal nerve, 854-5. (835)
Gluteus minimus, 374-5 ; Fig. 290. (373-4)
 Origin. Insertion. Structure. Nerve-supply. Action. Relations. Variations.

Disarticulate the pelvis and lower extremity from the trunk, dividing the spine between the third and fourth lumbar vertebræ. Divide the pelvis, cutting one-half an inch to the left of the median line in front and behind.

Pyriformis, 375-6-7 ; Fig. 288. (374)
 (Origin.) Insertion. Structure. Nerve-supply. Action. Relations. Variations.
Sciatic artery, 607-8 ; Fig. 369. (595-6)
 Branches of the sciatic artery :—
  Intrapelvic branches.
  Extrapelvic branches :—
   Coccygeal. Inferior gluteal. Muscular branches.
   Anastomotic branch, crucial anastomosis.
   Articular branches. Cutaneous branches.
   Comes nervi ischiatici.
Small sciatic nerve, 856-7 ; Fig. 467. (837)
Great sciatic nerve, 858 ; Fig. 467. (838-9)
Internal pudic artery, 608. (596-7)
 As it crosses the spine of the ischium :—
   Branches of the artery.
 The pudic nerve accompanies the artery, 857. (838)
Obturator internus and gemelli, 377-8 ; Fig. 288. (374-5-6-7)
 Obturator internus :—
   Origin. Insertion. Structure. Nerve-supply.
 Gemellus superior :—
   Origin. Insertion. Structure. Nerve-supply.
 Gemellus inferior :—
   Origin. Insertion. Structure. Nerve-supply.
   Action. Relations. Variations.
Quadratus femoris, 378-9 ;. Fig. 288. (377)
 Origin. Insertion. Structure. Nerve-supply. Action. Relations. Variations.
Obturator externus, 383-4. (382)
 Insertion. (The muscle will be shown later.)

The parietal structures of the pelvic cavity not already fully exposed, should now be demonstrated.

Pyriformis, 375. (374)
    Origin.   Insertion.   Structure.

Popliteal Space, 1225–6–7–8.   (1200–1–2–3–4)

The popliteal space is the lozenge-shaped space or hollow placed behind the knee-joint.

    Hollow of this space.
    Popliteal tendons.   Bursæ.
    Popliteal vessels :—
        The artery.   Popliteal vein.
        Superior articular arteries ; inferior.
        The external saphenous vein.
        Popliteal glands.

Expose the popliteal space by a median incision extending from a point six inches above, to one four inches below the joint ; make transverse incisions at each end of the median incision. Turn the skin each way from the median incision, exposing the superficial fascia ; in the superficial fascia will be found terminal branches of the small sciatic nerve and the upper portion of the internal saphenous vein.

Branches of small sciatic nerve, 856–7.   (837)
Short or external saphenous vein, 685.   (669–71)
Deep fascia of the leg, 390.   (388)

Turn off the deep fascia in the same manner as the integument, carefully remove the fat and connective tissue from the space, exposing the muscular boundaries and the structures in the space.

The muscular boundaries of the popliteal space are :—

    Above—Externally : Biceps—384—.   (382)
        Internally : Semi tendinosus—385—.   (383)   Semi-membranosus—386—(384) Gracilis—383—.   (381)   Sartorius (369)—368—Tendon of adductor magnus—382—.   (380)
    Below—Externally : Plantaris—393—, (391) and outer head of gastrocnemius.
        Internally : Inner head of gastrocnemius—391—.   (389)

At the upper end of the popliteal space the great sciatic nerve divides into two large trunks, the internal and the external popliteal nerves.

The internal popliteal nerve, 861.   (841)
    Cutaneous branch or nervus communicans tibialis.
    Muscular branches.
    Articular branches.
The external popliteal nerve, 859.   (839)
    Cutaneous branches.   Nervus communicans peronei.
    Articular branches.   Recurrent articular nerve.
The popliteal vein, 686.   (671)

Chief variations of the popliteal.

Popliteal artery, 621–2–3.   (609–10–11)
    Relations :—In front.   Behind.   Inner side.   Outer side.

Principal variations in the popliteal.

    Branches of the popliteal artery, 623–4–5.   (612–13)
        Cutaneous branches.
        Muscular or sural branches :—
            Upper muscular or superior sural.
            Inferior muscular or sural branches.
        Articular :—
            Superior external articular.
            Superior internal articular.
            Inferior internal articular.
            Inferior external articular.
            Azygos articular.
Geniculate branch of the obturator nerve, 850.   (831)
    Popliteal glands, 707.   (691)

Back of the Thigh.

Continue the median incision on the back of the thigh and turn off the skin. In the superficial fascia will be found cutaneous branches of: The small sciatic, 856-7 (837); External cutaneous, 851 (832); Internal cutaneous, 852 (832-3); Obturator, 849 (829); these should be exposed in place, then turn off the deep fascia (364) and expose the structures underneath.

The hamstring muscles, 384 ; Fig. 291. (382)
    Flexor biceps femoris, 384-5. (382-3)
        Origin, long head, short head. Insertion.
        Structure. Nerve-supply. Action. Relations. Variations.
    Semi-tendinosus, 385. (383-4)
        Origin. Insertion. Structure. Nerve-supply. Action. Relations.
    Semi-membranosus, 386-7. (384-5)
        Origin. Insertion. Structure. Nerve-supply. Action. Relations. Variations.

Note the bursæ found in connection with the popliteal tendons, 1226-7.

Great sciatic nerve. External and internal popliteal nerves, 858-9. (838-9)
    Branches.
Terminal branches of the perforating arteries, i-ii-iii-iv, 620. (608-9)

Divide the adductor magnus and quadratus femoris at their origin and expose the obturator externus. Carefully raise the obturator externus and trace the obturator vessels and nerve. Note the insertion of the psoas and iliacus.

Obturator externus, 383-4. (382)
    Origin. Insertion. Structure. Nerve-supply. Action. Relations.
Obturator artery, 606-7. (595)
    External branch. Internal branch.
Obturator nerve, 849. (829-30)
    Twigs from anterior and posterior branches to hip joint, and to muscles.
Psoas magnus, 366-7. (365-6-7)
    Insertion. Action. Relations in the thigh.
Iliacus, 368. (367)
    Insertion. Action. Relations.
Note:—Muscles in relation with the hip joint, 264. (268)
        In front. Above. Above and behind. Below and behind. Internally.

LANDMARKS OF THE LEG, ANKLE, AND FOOT.

The Leg, 1228-1233. (1204-1209)
Bony landmarks.
Muscular prominences.
Vessels :—
    Saphena veins ; internal, external.
    Popliteal artery.
    Line of posterior tibial artery.
    Line of anterior tibial artery.

The Ankle, 1233. (1209)
Bony landmarks :—
    Malleoli ; internal, external.

The Foot, 1238. (1214)
Bony landmarks :—
    (A) Along the inner aspect of the foot.
    (B) Along the outer aspect.
Levels of the joints and lines of operations, 1238-9-40. (1214-15-16)
        Syme's amputation.
        Pirogoff's amputation.
        Chopart's medio-tarsal amputation.
        Lisfranc's or Hey's or tarso-metatarsal amputation.
        Amputation of the great toes.
Line of dorsal artery, 1242. (1218)
Line of plantar arteries, 1242. (1218)

Cutaneous nerves, 1242. (1218)
    Musculo-cutaneous nerve.
    Anterior tibial.
    External saphenous nerve.
    Internal saphenous nerve.

Dissection of Front of Leg and Dorsal Surface of Foot.

Continue the median incision down the leg, in front of the ankle and over the dorsum of the foot to the second toe. Make transverse incisions at the middle of the leg, at the ankle, and at the base of the toes, also a median incision on the dorsum of each toe, to the end. Reflect the skin each way from the median incisions and expose the superficial fascia, veins, and cutaneous nerves.

Superficial veins of the lower limb, 683–5 ; Fig. 397. (669)
    Long or internal saphenous vein.
Superficial lymphatic vessels, 705 ; Fig. 400. (689)
Cutaneous nerves, Fig. 464 ; P. 851. (831)
    Long or internal saphenous nerve, 852–3. (833)
    Cutaneous branches of the external popliteal, 859. (839)
    Branches of the musculo-cutaneous, 859. (839)
        Internal branch. External branch.
    Internal branch of the anterior tibial, 861. (841)
    External or short saphenous, 861. (841)

Turn off the superficial fascia and expose the deep fascia of the leg.

Deep fascia of the leg, 390. (388)
Anterior annular ligament, 1237 (1213); also, 390–1. (388)
    Upper, above the level of the ankle-joint.
    Lower, over the ankle-joint.

Turn off the deep fascia, leaving the annular ligament over the ankle-joint, to retain the tendons in place. As each muscle is exposed, the tendon should be traced to its terminal insertion.

Muscles of the front of the leg, 410 ; Figs. 297–8. (407)
Tibialis anticus, 410. (407)
    Origin. Insertion. Structure. Nerve-supply. Action. Relations. Variations.
Extensor longus digitorum, 412–13. (409–10)
    Origin. Insertion. Structure. Nerve-supply. Action. Relations. Variations.
Peroneus tertius, 413–14. (410)
    Origin. Insertion. Structure. Nerve-supply. Action. Relations. Variations.
Extensor proprius hallucis, 410–11–12. (407–8–9)
    Origin. Insertion. Structure. Nerve-supply. Action. Relations. Variations.
Tendons in front of ankle, 1233–4. (1209–10)
Bursæ and synovial membranes, 1241. (1217)
Extensor brevis digitorum, 415. (411–12)
    Origin. Insertion. Structure. Nerve-supply. Action. Relations. Variations.
Anterior tibial nerve, 859–60–1 ; Fig. 468. (839–40–1)
    Branches :—Internal branch.
              External branch.
Musculo-cutaneous nerve, 859 ; Fig. 468. (839)
    External branch.
    Internal branch.
Anterior tibial artery, course of the vessel, 632–3–4 ; Fig. 380. (620–1–2)
    Relations :—Posteriorly. Inner side. Outer side. In front.
    Branches of the anterior tibial artery :—
            Anterior tibial recurrent.
            Muscular branches.
            Internal malleolar.
            External malleolar.
Dorsalis pedis artery, 634–6–7 ; Fig. 380. (622–4–5)
    Relations :—Behind. In front. Outer side. Inner side.
    Branches of the dorsalis pedis artery, 636. (624)
        Tarsal branches :—
            Internal tarsal.
            External tarsal.

Metatarsal artery :—
    Dorsal interosseous arteries.
        Dorsal digital vessels.
Dorsalis hallucis :—
    Dorsal digital branches.
Communicating.
Anterior peroneal artery, 627; Fig. 380. (615)
Deep veins—of the lower extremity, 683 and 685. (669 and 671)
Deep lymphatic vessels of the lower limb, 707. (691)

Peroneal Region.

Turn off the deep fascia covering the peronei muscles, leave enough of the external annular ligament to retain the tendons in place as they pass under it.

External annular ligament, 391 and 1237. (388 and 1213)
Muscles on the outer side of the leg, 415 ; Figs. 292, 297. (412)
Peroneus longus, 415–16. (412–13)
    Origin. (Insertion.) Structure. Nerve-supply. Action. Relations. Variations.
Peroneus brevis, 417. (413–14)
    Origin. (Insertion.) Structure. Nerve-supply. Action. Relations. Variations.
External popliteal nerve, 859. (839)
    Recurrent articular nerve.
    Musculo-cutaneous nerve.
    Anterior tibial nerve.

Posterior Tibio-Fibular Region.

Note the cutaneous nerve-supply, Fig. 464 ; P. 851. (831)

Continue the median incision to the middle of the heel. Make a transverse incision from the internal to the external malleolus, passing under the middle of the heel. Reflect the skin from the posterior surface of the leg and from the posterior portion of the heel, exposing the superficial fascia, veins, and nerves.

Superficial veins of the lower limb, 683–5. (669–71)
    Branches of internal saphenous vein.
    The short or external saphenous vein.
Cutaneous nerves, Fig. 464 ; P. 851. (831)
    Posterior branch of the internal cutaneous nerve, 852. (832–3)
    Internal saphenous nerve, 852–3. (833)
    Internal calcanean or calcaneo-plantar cutaneous branch, 863. (843)
    Terminal branches—of the small sciatic nerve, 856–7. (837–8)
    External or short saphenous nerve, varieties, 861. (841)
Deep fascia of the leg, 390. (388)
    Internal annular ligament, 391 and 1237. (388 and 1213)

Divide the deep fascia in the median line and turn out and in from this incision ; preserve a portion of the internal annular ligament.

Gastrocnemius, 391–3 ; Fig. 291. (389)
    Origin :—Outer head ; inner head. Insertion.
    Structure. Nerve-supply. Relations. Variations.

Divide the gastrocnemius transversely about three inches above the tendo-Achillis and raise the upper portion to expose the structures underneath.

Plantaris, 393. (391)
    Origin. Insertion. Structure. Nerve-supply. Action. Relations. Variations.
Soleus, 394–5–6. (392–3)
    Origin. Insertion. Structure, tendo-Achillis (behind the ankle, 1235). (1210)
    Nerve-supply. Action. Variations.

Divide the soleus transversely in the same manner as the gastrocnemius and raise to expose the structures underneath, then divide the upper portion of the muscle vertically in the median line, note the tendinous arch of origin (3) between the tibia and fibula.

Deep tibial fascia, 396. (393)
Popliteal artery; terminal branches, 625; Fig. 374. (613)
    Anterior tibial artery, 632. (620)
        Posterior tibial recurrent, 634. (620)
    Posterior tibial artery, 626–7; Fig. 377. (614–15)
        Relations:—Anteriorly. Posteriorly. At the inner ankle.
        Branches, 628. (615–16)
            Muscular branches. Medullary artery. Cutaneous branches.
            Communicating branch.
            Malleolar or internal malleolar branches.
            Calcanean or internal calcanean branch.
Vessels divided in amputation, 1233. (1209)
Posterior tibial nerve, 862–3. (842–3)
    Branches:—
        Internal calcanean or calcaneo-plantar branch.
Deep lymphatic vessels of the lower limb, 707. (691)
Deep lymphatic glands of the lower extremity, 707. (691)
Popliteus, 394; Fig. 292. (391–2)
    Origin. Insertion. Structure. Nerve-supply. Action. Relations. Variations.

Expose and separate the deep muscles, follow the tendons to the annular ligament; the insertion will
be shown later in the dissection of the foot.

Flexor longus hallucis, 398; Fig. 292. (395–6)
    Origin. (Insertion.) Structure. Nerve-supply. Action. Relations. Vari-
    ations.
Flexor longus digitorum, 396–7; Fig. 292. (393-4-5)
    Origin. (Insertion.) Structure. Nerve-supply. Action. Relations. Vari-
    ations.
Tibialis posticus, 398–9; Fig. 292. (396)
    Origin. (Insertion.) Structure. Nerve-supply. Action. Relations.
Peroneal artery, 627–8; Fig. 377. (615–16)
    Branches of the peroneal artery:—
        Muscular branches. Medullary. Communicating branches.
        Cutaneous branches. External calcanean. Terminal branch.
        Anterior peroneal artery, Fig. 380.
Points in tenotomy and guides to the tendons, 1237–8. (1213–14)
    Tendo-Achillis. Tibialis anticus. Tibialis posticus. Peronei.

Sole of the Foot.

Continue the median incision to the end of the middle toe. Make a transverse incision at the base of the
toes, and from this a median incision to the end of each toe, remove the skin, exposing the superficial
fascia and cutaneous nerves.

The sole of the foot is supplied by the internal calcanean or calcaneo-plantar cutaneous branch of the posterior tibial, and cutaneous branches from the internal plantar
and external plantar nerves, 863; Fig. 469. (843)

Dissect off the superficial fascia and expose the plantar fascia. Carefully preserve the superficial nerves
and vessels.

Plantar fascia, 399–400–1. (397–8)
    Central part. Inner portion. External portion.
    Superficial transverse ligament.

"The lateral portions of the plantar fascia should be raised from the subjacent muscles. The three
superficial muscles of the sole are then exposed to view and their connections can be studied. The
flexor brevis digitorum is placed in the middle, the abductor minimi digiti extends along the outer margin of the sole, and the abductor hallucis along the inner margin of the sole. In the interval between
the abductor hallucis and flexor brevis digitorum the internal plantar nerve and artery will be found.
Follow the nerve toward the toes and dissect out its four digital branches. In doing so, care must be
taken of the muscular twigs which are given to the flexor brevis hallucis and the innermost lumbrical
muscle. Slender branches of the internal plantar artery accompany the digital nerves. Now trace the
trunk of the internal plantar nerve backward by carefully separating the flexor brevis digitorum and
the adductor hallucis along the line of the intermuscular septum. It will be found to give a branch of
supply to each of these muscles. In the next place, separate the contiguous borders of the flexor brevis

digitorum and abductor minimi digiti. The external plantar artery and nerve lie for a short portion of their course in the interval between these muscles. Approaching the prominent base of the fifth metatarsal bone, the artery disappears from view by turning inward under cover of the flexor tendon. At the same point the external plantar nerve divides into its superficial and deep divisions. The deep division of the external plantar nerve cannot be dissected at present, as it accompanies the external plantar artery. The superficial division, however, should be traced to its distribution." (Cunningham.)

Flexor brevis digitorum or flexor perforatus, 401–2 ; Fig. 293. (399)
 Origin. Insertion. Structure. Nerve-supply. Action. Relations. Variations.
Abductor hallucis, 401 ; Fig. 293. (398–9)
 Origin :—Outer head ; Inner head. Insertion. Structure.
 Nerve-supply. Action. Relations. Variations.
Abductor minimi digiti, 402–3 ; Fig. 293. (399–400)
 Origin. Insertion. Structure. Nerve-supply. Action. Relations. Variations.

Divide the muscles of the superficial group near their origin from the os calcis and turn them forward ; note the arterial and nerve branches to each, expose the superficial portion of the plantar arteries and the plantar nerves.

Internal plantar artery, 631–2 ; Fig. 378. (619–20)
 Relations. Branches :—Muscular branches. Cutaneous branches.
       Articular. Anastomotic. Superficial digital.
External plantar artery (superficial portion), 628–9 ; Fig. 378. (616–17)
 Relations in the first part of its course, 629–30. (617)
 Branches of the external plantar artery : —
  Muscular branches. Calcanean. Cutaneous. Anastomotic.
The veins of the foot and leg, 685. (671)
Internal plantar nerve ; cutaneous branches, 863 ; Fig. 469. (843)
 First digital branch. Second branch. Third branch. Fourth branch.
External plantar nerve, 863 ; Fig. 469. (843)
 Superficial division :—
  Internal branch.
  External branch.

Expose the tendon of the flexor longus digitorum, and trace forward each division to its terminal insertion.

Flexor longus digitorum, 396 ; Fig. 294. (393–4–5)
 Insertion. Structure. Action.
Flexor accessorius digitorum pedis, 404 ; Fig. 294. (400–1)
 Origin :—Inner head ; Outer head. Insertion. Structure.
 Nerve-supply. Action. Relations. Variations.
The four lumbricales, 404–5 ; Fig. 294. (401–2)
 Origin. Insertion. Structure. Nerve-supply. Action. Relations.
Flexor longus hallucis, 398 ; Fig. 294. (395)
 Insertion. Structure. Action.

To expose the next layer of muscles, divide the flexor accessorius at its origin, cut the tendons of the flexor longus digitorum and the flexor longus hallucis in front of the annular ligament and turn forward to the toes. Leave the plantar vessels and nerves in place.

Flexor brevis hallucis, 405–6 ; Figs. 294–5. (402–3)
 Origin. Insertion. Structure. Nerve-supply. Action. Relations. Variations.
Adductor hallucis, 406–7 ; Fig. 295. (403–4)
 Origin. Insertion. Structure. Nerve-supply. Action. Relations. Variations.
Transversus pedis, 407 ; Fig. 295. (404)
 Origin. Insertion. Structure. Nerve-supply. Action. Relations. Variations.
Flexor brevis minimi digiti, 407–8 ; Fig. 295. (404–5)
 Origin. Insertion. Structure. Nerve-supply. Action. Relations. Variations.

Divide the flexor brevis and the adductor hallucis at their origin, and throw them forward to expose the plantar arch and the deep branch of the external plantar nerve.

Plantar artery ; second part of its course, 630–1 ; Fig. 379. (618–19)
 Branches :—Articular.
    Digital or plantar digital :—
     Collateral digital arteries.

Fifth plantar digital or princeps hallucis, 637. (625)
Posterior perforating, 630. (618)
Deep division of the external plantar nerve, 863. (843)
The heads of the metatarsal bones are connected on their plantar aspect by the transverse ligament, 293. (296)

Divide the transverse ligament, separate the toes and expose the interosseous muscles. Note the communication between the dorsal and plantar arteries.

The interossei :—Plantar ; Dorsal, 408–9 ; Fig. 296. (405–6)
    Plantar interossei :—
        Origin. Insertion.
    Dorsal interossei :—Insertion.
    Structure. Nerve-supply.
    Action of :—Plantar interossei.
              Dorsal interossei.
    Relations.

Trace the tendons and demonstrate the insertion of the tibialis posticus ; peroneus brevis ; and peroneus longus.

Tendons on inner side of ankle, 1235–6–7. (1211–12–13)
    Tibialis posticus, 399. (396)
        Insertion. Action.
Tendons at outer ankle, 1237. (1213)
    Peroneus brevis, 417. (413)
        Insertion. Action.
    Peroneus longus, 415–16. (412–13)
        Insertion. Structure. Action.

Demonstrate the anastomosis about the knee-joint, Fig. 376, P. 625 (613), also about the ankle-joint, and foot, Fig. 381, P. 635. (623)

Remove the muscles from the pelvis and lower extremity, leaving about four inches of each flexor tendon at the knee-joint ; demonstrate the ligaments of each joint as indicated.

The Articulations, 186–7. (195–6)

Bones.
Cartilage :—
    Articular ; hyaline variety.
    White fibrous :—
        As interarticular cartilage.
        As circumferential.
        As connecting fibro-cartilage.
The ligaments.
Synovial membrane.

The Various Kinds of Articulations, 187–8. (196–7)

Synarthrosis :—
    True sutures.
    False sutures.
    Grooved sutures.
Amphiarthrosis.
Diarthrosis :—
    Arthrodia.
    Ginglymus, ginglymo-arthrodial.
Enarthrodia. Trochoïdes.
Table of the various classes of joints.

The Various Movements of Joints, 188–9. (197–8)

    Gliding. Angular. Rotation. Circumduction.

THE ARTICULATION OF THE PELVIS WITH THE SPINE, 209-10-11. (216-17-18)

Sacro-lumbar ligament.
Ilio-lumbar ligament.
Arterial supply. Nerve-supply.
Movements.

ARTICULATIONS OF THE PELVIS, 211; Figs. 206 to 213. (218)

(a) Sacro-iliac.
(b) Sacro-coccygeal.
(c) Intercoccygeal.
(d) Symphysis pubis.

(a) Sacro-iliac synchondrosis and sacro-sciatic ligaments, 211-12-13-14. (218-19-20-21)
    Class—*Amphiarthrosis.*
    Anterior sacro-iliac ligament.
    Superior sacro-iliac ligament.
    Posterior sacro-iliac ligament.
    Inferior sacro iliac ligament.
    Interosseous ligament.
        Ear-shaped cartilaginous plate.
        Great or posterior sacro-sciatic ligament. Falciform process.
    Small sacro-sciatic ligament.
        Arterial supply. Nerve-supply. Movements.

(b) Sacro-coccygeal articulation, 214-15-16. (221-2-3)
    Class—*Amphiarthrosis.*
    Anterior sacro-coccygeal ligament.
    Posterior sacro-coccygeal ligament.
    Supracornual ligament.
    Intertransverse ligament.
    Intervertebral substance.
        Arterial supply. Nerves. Movements.

(c) Intercoccygeal Joints, 216. (223)
(d) The Symphysis Pubis, 216-17-18. (223-4-5)
    Class—*Amphiarthrosis.*
    Superior ligament.
    Posterior ligament.
    Anterior ligament.
    Inferior or infrapubic ligament.
    Interosseous cartilage.
        Arterial supply. Nerve-supply. Movements.

THE ARTICULATIONS OF THE LOWER LIMB, 258. (263)

1, 2, 3, 4, 5, 6, 7, 8, 9.

1. THE HIP JOINT, 259 to 266; Figs. 237 to 244. (263 to 270)

    Class—*Diarthrosis.* Subdivision—*Enarthrodia.*
    Capsular ligament, at pelvis, at femur. Retinacula.
        Ilio-femoral.
        Ischio-femoral band.
        Pectineo-femoral band.
        Tendino-trochanteric band.

Open the capsular ligament, demonstrate the ilio-femoral band, then divide it.

    Ligamentum teres.
    Transverse ligament.
    Cotyloid fibro-cartilage.
    Synovial membrane.
        Arterial supply. Nerve-supply.
        Movements.

2. THE KNEE-JOINT, 266 to 277; Figs. 245 to 252. (271 to 280)

    Class—*Diarthrosis.* Subdivision—*Ginglymus.*
    External ligaments :—
        The fibrous expansion of the extensor tendons.
        Ligamentum patellæ.
        Internal lateral ligament.
        External lateral ligament.
        Posterior lateral ligament, or ligamentum Winslowii.
        Capsular or anterior ligament.

Open the joint in front, cutting the ligaments transversely above the patella. Demonstrate the internal ligaments and the cartilages.

9

Internal ligaments :—
    Anterior crucial ligament.
    Posterior crucial ligament.
    Interarticular or semilunar fibro-cartilages.
        External semilunar cartilage.
        Internal semilunar cartilage.
    Transverse ligament.
    Coronary ligaments.
    Synovial membrane : (also, 1223-4). (1199-1200)
        Ligamentum mucosum. Alar ligaments.
    Arterial supply.
    Nerve-supply :—
        Great sciatic. Anterior crural. Obturator.
    Movements :—
        Antero posterior spiral curve.
        Extension. Flexion. Pronation.
        Supination. Sliding movements.

3. THE TIBIO-FIBULAR UNION, 277-8-9. (281-2-3)
    (a) Superior tibio-fibular joint.
    (b) Middle tibio-fibular union.
    (c) Inferior tibio-fibular joint.
    (a) Superior tibio-fibular joint.

    Class—*Diarthrosis*. Subdivision—*Arthrodia*.
    Capsular ligament.
    Anterior tibio-fibular ligament.
    Posterior tibio-fibular ligament.
    Superior interosseous ligament.
    Biceps tendon.
    Synovial membrane.
    Arterial supply. Nerve-supply. Movements.

    (b) Middle tibio fibular union.

    The interosseous membrane.

    (c) Inferior tibio-fibular articulation.

    Class—*Diarthrosis*. Subdivision—*Arthrodia*.
    Anterior inferior tibio-fibular ligament.
    Posterior inferior tibio-fibular ligament.
    Inferior interosseous ligament.
    Transverse ligament.
    Synovial membrane.
    Nerve-supply. Movements.

4. THE ANKLE-JOINT, 279-80-1-2 ; Figs. 253-4-5. (283-4-5)
    Class—*Diarthrosis*. Subdivision—*Ginglymus*.
    Anterior ligament.
    Posterior ligament.
    Internal lateral or deltoid ligament.
    External lateral ligament :—
        Anterior fasciculus. Middle fasciculus. Posterior fasciculus.
    Synovial membrane.
    Nerve-supply. Arterial supply.
    Movements.

5. THE TARSAL JOINTS, 282 to 290 ; Figs. 253 to 258. (286 to 293)
    (a) Calcaneo-astragaloid union.
    (b) Articulations of the anterior portion of the tarsus.
    (c) Medio-tarsal joint.
        Synovial membranes in tarsus, 1241-2. (1217-18)
    (a) Calcaneo-astragaloid union.

    Class—*Diarthrosis*. Subdivision—*Arthrodia*.
    (i) Posterior calcaneo-astragaloid joint.
        Interosseous ligament, anterior interosseous ligament.
        External calcaneo-astragaloid ligament.
        Posterior calcaneo-astragaloid ligament.
        Internal calcaneo-astragaloid ligament.
        Synovial sac.
        Nerve-supply. Arteries.

(ii) Anterior calcaneo-astragaloid joint.
    Interosseous ligament.
    Antero-internal calcaneo-astragaloid.
    External calcaneo-scaphoid.
    Movements.
    Synovial membrane.
    Arteries. Nerves.

(b) The articulations of the anterior part of the tarsus.

    Class—*Diarthrosis.* Subdivision—*Arthrodia.*

    (i) Cubo-scaphoid union.
        Dorsal cubo-scaphoid ligament.
        Plantar cubo-scaphoid ligament.
        Interosseous cubo-scaphoid ligament.

    (ii) Scapho-cuneiform articulation.
        Dorsal scapho-cuneiform ligament.
        Internal scapho-cuneiform ligament.
        Plantar scapho-cuneiform ligament.

    (iii) Intercuneiform and (iv) cubo-cuneiform articulations.
        Dorsal ligaments.
        Plantar ligaments.
        Interosseous ligaments.
        Arterial supply. Nerves. Movement.

(c) The medio-tarsal or transverse tarsal joints.

    (i) Astragalo-scaphoid articulation.
        External calcaneo-scaphoid ligament.
        Inferior calcaneo-scaphoid ligament.
        Astragalo-scaphoid ligament.
        Synovial membrane.

    (ii) Calcaneo-cuboid articulation.
    Class—*Diarthrosis.* Subdivision—*Arthrodia.*
        Internal or interosseous calcaneo-cuboid ligament.
        Dorsal calcaneo-cuboid ligament.
        Long inferior calcaneo-cuboid.
        Short inferior calcaneo-cuboid.
        Synovial membrane.
        Arterial supply. Nerve-supply.
        Movements, lateral motion.

6. TARSO-METATARSAL ARTICULATIONS, 290-1-2. (293-4-5)

    Class—*Diarthrosis.* Subdivision—*Arthrodia.*
    Three articulations—(a), (b), (c).
        Arteries for tarso-metatarsal joints.
        Nerve-supply. Movements.

    (a) Internal tarso-metatarsal joint.

    (b) Middle tarso-metatarsal joint.

        The dorsal ligaments.
        The plantar ligaments.
        The interosseous ligaments.
        Synovial membrane.

    (c) Cubo-metatarsal joint.
        Plantar cubo-metatarsal ligament.
            External cubo-metatarsal ligament.
        Dorsal cubo-metatarsal ligament.
        Interosseous ligament.
        Synovial membrane.

7. INTERMETATARSAL ARTICULATIONS.

    Dorsal ligaments.
    Plantar ligaments.
    Interosseous ligaments.
    Synovial membrane.
    Arterial and nerve-supply.
    Movements.

    The union of the heads of the metatarsal bones.
        Transverse ligament.